TO Robert
member of The Odyssean
Club, but also a very informed
and dear friend; indeed, one of
the better persons I have ever
had the pleasure to meet and know.

                    Ted Pryor

Wealth Building Lessons of Booker T. Washington

for a New Black America

T.M. Pryor

Copyright © April 1995 by Theadore M. Pryor

We gratefully acknowledge permission to reprint brief quote from *W.E.B. BuBois: Biography of a Race, 1868-1919* by David L. Lewis © 1993, Henry Holt & Company; Quote from *Booker T. Washington and the Negro Place in American Life* by Samuel R. Spencer, Jr. © 1955, Little Brown & Company.

Duncan & Duncan, Inc., <u>Publishers</u>
Mailing Address: P.O. Box 1137, Edgewood, MD 21040
Phone: 410-538-5580 Fax: 410-538-5584

Library of Congress Catalog Card Number: 95-67018

Pryor, Theadore M., 1921–
 Wealth Building Lessons of Booker T. Washington
1. Black Businesses. 2. Booker T. Washington 3. Afro-American History 4. Afro-Americans, economic empowerment of, 5. W.E.B. DuBois 6. Race relations in America 7. Afro-Americans-economic conditions

ISBN 1-878647-21-0

All rights reserved. No part of this book may be reproduced or transmitted (except for brief excerpts for book reviews and articles) in any form or by any means, electronic or otherwise, without the permission in writing from the publisher.

9 8 7 6 5 4 3 2 1

Dedication

In fond memory of Amber whose thirty-five days with us were not long enough for her to call me "Popi" like my other grandchildren Meiyen, Martene, Nikole, Erin and Alexis to whom this book is dedicated.

Acknowledgments

In 1967, on a trip to New York in search of a record titled "Long Gone," a blues instrumental by Sonny Thompson recorded in the late 1940's, I happened to tell the West Indian clerk, who waited on me, that I had come 115 miles from Hartford, Connecticut. I was there trying to find a replacement copy for the record one of my four children had accidentally broken.

The clerk lit up at the name Connecticut. It brought to mind, he said, Venture Smith, a heroic figure of the American slave system. The clerk said Smith offered fierce resistance to being whipped. Anyone who tried, Smith's master included, was in for a rude awakening, the clerk told me. Smith even defended his wife when the mistress tried to flay her. Though intrigued by the story, I was mildly resentful that a black foreigner had as a hero an American slave whom I was sure was unknown to nearly all American blacks. So while I never learned the clerk's name, and I did not find the record "Long Gone," I am indebted to him for introducing me to Venture Smith as a significant black historical figure. This meeting

Acknowledgement

caused me to read black history-related publications extensively for the first time ever.

While it was several years before I found anything on Venture Smith, I came across a plethora of material on Booker T. Washington and William E. B. DuBois. I was particularly disturbed by Langston Hughes's treatment of the two men in his publication, *A Pictorial History of the Negro in America*, which I first read about 1970. I thought Hughes evidenced veiled derision and contempt for Booker T. Washington vis-a-vis his treatment of DuBois, as though DuBois had rescued American blacks from Washington's deadly enchantment.

Hughes's account was at variance with what I understood, and it did not square with the upbeat, positive and refreshing stories the late Berkeley G. Burrell, then president of the National Business League, often related about Washington. It was because of Hughes's unfortunate, and it seemed, very partial treatment or mistreatment of Washington, that I began an extensive study of the two men and began paying closer attention to how other writers handled the two black historical figures. So in a oddballish way, I am grateful to Hughes for awakening in me a more than passive interest in black history.

When I retired in 1981, pioneer space medicine physician, the late Dr. Vance Marchbanks and later my very dear friend, Morehouse College professor Dr. William G. Pickens, insisted that in my role as leader of the Hartford black business establishment, there were some lessons learned that should be told. I was encouraged to tell these lessons in a book since black business leadership for a black community the size of Hartford's was a first. To the extent that I could, I have done so here. So I am grateful to Dr. Marchbanks and Dr. Pickens for their urgings.

Professor Pickens and my good friend, Attorney Arnold

Sidman, volunteered as critics of my work and spent hours at lunch and in their offices in critical discussion of issues I planned to include in the manuscript. Without their advice and counsel, I might have given up. Dr. Robert Brisbane, professor emeritus of Morehouse College and Dr. Robert Michael Franklin of Emory University, both published authors, and Dr. Asa Atkins, former medical director, who is a friend of some of Booker T. Washington's great grandchildren, were essential boosters during the writing of this book.

Of course, my wife, Sophornia, was excessively supportive and helpful. She smiled through the several times I canceled a planned outing with her in favor of a trip to the library to do research or to work on a new perception to be included in the book. Her patience and understanding made writing the manuscript a most satisfying experience.

Contents

Introduction	8
One: Before the Speech: Message From the Grave	13
Two: Booker T. Washington— An Uncommon Perspective	33
Three: Questions and Answers: Part 1	51
Four: Questions and Answers: Part 2	77
Five: A Capital Accumulation Plan for Black Businesses	108
Six: Where Is the Black Press?	122
Epilogue	128
Appendix: Booker T. Washington's Cotton States Exposition Speech	133
Select Bibliography	139
Index	142

Introduction

The Establishment will escalate blacks up to the mezzanine floor or higher. It will elevate them to the topmost floors, and it will airlift them up to the plane's cruising altitude, but it will not lift them one basis point up the ladder of power. The heftiness for that hoist must come from blacks themselves.

That was one of the intrinsic lessons I gleaned from my very first reading of Booker T. Washington's erudite autobiography *Up From Slavery*. **I was able to figure out too that the business establishment was a capitalistic society's economic base and that the economic base and power are in direct proportion.**

As a black septuagenarian, I am a black elder, a status which carries with it a responsibility. In his novel, *The Chosen*, Chaim Potok tells what the responsibility is. In a 1930's New York setting, one Jewish character informs another that the Nazis in Germany are killing the Jewish elders. The

Introduction

other asks if the elders are being killed off, who will be left to water the roots? If the reader understands "water" to mean teach and "roots" to mean younger generations, then I am a teacher to succeeding generations of my people. My view of Booker T. Washington differs considerably from the usual denigration of him by Washington-phobes who currently seem to be published most.

With the 100th anniversary of Washington's legendary 1895 Atlanta Cotton States Exposition Speech being upon us, American blacks, non-blacks and the world should have a range of views of Washington so that each person can make a personal determination of his historical measure.

I charge that American black history for odd and sundry motives is now revisionist history for two reasons. First, since 1865, American blacks have been exalted by too many black writers as an heroic people when, in fact, we were mostly a grossly maligned ethnic group (caused by degrading socialization policies of the government and the majority racial group). Even so, blacks survived and even grew despite scores of years of passivity that for a while seemed to be innate.

Secondly, there is a propensity of black writers to anoint and elevate to sainthood individuals with no record of substantive accomplishments. In this regard, I contend Malcolm X has been oversubscribed, probably at the expense of militant black editor T. Thomas Fortune, who was advisor to Booker T. Washington. Martin Luther King, Jr. has been grossly undersubscribed and the surface of his accomplishments barely scratched and shallowly analyzed, just as has been the Montgomery Movement he captained.

A. Philip Randolph is undersubscribed as will be shown. William Edward Burghardt DuBois has been disproportionately oversubscribed as I shall attempt to show in Chapter

Four. Booker T. Washington has been disgracefully and unresponsibly undersubscribed and inexcusably defamed as I hope to prove throughout this book.

The areas cited above as having been historically misplayed by black (and white) writers will not necessarily be highlighted in this book but will be treated in normal fashion. However, I hope that the reader will note where my version varies with conventional boilerplate versions and pause to reflect.

One of my previous articles was published in the 1993 Fall issue of the *Boule Journal*, the official publication of Sigma Pi Phi Fraternity. The article (partly contained in Chapter Two) caused a myriad of questions and comments from readers and gave me additional incentive to complete this book. I have included answers to many of these challenging questions in Chapters Three and Four.

My answers and comments show me at wide variance with commonly-held perceptions. For example, I disagree that DuBois founded the NAACP. I am contemptuous of the song "Lift Every Voice and Sing" for not-too-obvious reasons. I accuse blacks of making disproportionately large the minuscule—like labelling as a movement many insignificant activities. And, I insist that the history-worthiness of anyone should be proportionate with how one's activities impacted a **significant number of the general population during his time and how comprehensive, definitive, useful and relevant his legacy is now.**

To encourage widespread and easy reading, I have avoided making this a "scholarly" work; rather I preferred to draw heavily on what I could recall of what I have read that relates to the subject and occasionally citing a reference in the text. All-in-all, I hope to accomplish six things with the publication of this book?

Introduction

1. To stir in American blacks an interest in their history as non-slaves, giving careful attention to the role their leadership asked each person to play and how each person acted out that role.
2. I wish for an understanding of the indispensable dominant heroic trio (Chapter Two) perception of blacks and the world for an appreciation of their role in the making of America.
3. I hope for blacks' curiosity to be piqued concerning the necessity for total organization of black businesses as a precondition for the black community to make measurable progress and to ask questions about the progress being made and also to develop a method for determining who is helping and who is hindering the organizational process.
4. I wish to stir up a black clamor for the erection in Washington of a national black monument as black America's bonding symbol underwritten entirely by blacks.
5. I wish for the Full Personhood Principle, introduced in Chapter Four, to become every black child's personal affirmation of allegiance to self and attainment of full personhood to become that child's most cherished ambition.
6. Finally, I wish to arouse the concern of the American black business establishment in the status of black America's oldest black-founded organization, the National Business League (NBL), to the extent that a NBL Restoration and Preservation Committee be formed to address the possibility of perpetually continuing the NBL.

Washington's teachings and philosophy for moving the black community into a growth attitude were not new, but taken from history—that is, little common sense lessons learned from life. The talking heads of Washington's era preached and emulated the doctrine to get attention rather than teach the dignity and nobility inherent in severe and purposeful striving that often lead to recognition. To this very day, talking heads prevail, hence another reason for this book as a change of pace.

All-in-all, the attempt in this book has been to develop in black Americans an understanding of our leadership whose legacy is substantive enough to be understood by the masses. For the **masses,** in the final analysis, **must provide the fuel for forward mobility of the American Black Community.**

T. M. Pryor

—ONE—

Before the Speech—Message From the Grave

If Booker T. Washington could return among us and be afforded an overview of the American Black Community, he would probably take stock of black businesses, his most urgent concern while he lived, and say with a tear-filled heart and in a choked voice: **"Negro businesses make no measurable contributions to the smog problem with which America now wrestles."** But, Washington was a practical man. He was not an idle dreamer. He saw things as they were and dealt with them from that perspective to mold them to his liking.

Washington founded and developed what today is Tuskegee University, which provided the facilities to allow George Washington Carver's rescue of the South's failing farm industry. Tuskegee enhanced the quality of other domestic farmlands and those of many countries beyond America's borders. It enriched the minds of thousands of black youths with ambitions for higher learning and usable industrial skills. Washington would take comfort in the progress made by blacks in education, statistics that are rarely noted when black achievements are recited to black youth.

In 1904, for instance, there were about 2,500 black college graduates in America or about 300 more than W. E. B. DuBois's 2,200 or so figure for 1899 in his 1903 essay, "The Talented Tenth." So, in the unlikely event there were zero black college graduates in 1865 when the American Black Community was born, then it had gained less than 100 college graduates a year during its thirty-nine years of existence.

In his book, *The Black Man in White America*, published in 1944, John C. Van Deusen estimated that in 1942 there were about 33,000 degree-bearing blacks in America or an average gain of slightly more than 800 a year since 1904. In 1992, the United States Census Bureau put the number slightly in excess of two million for an average yearly gain of about 40,000 black college graduates since 1942. And, while that is only about five percent of the black population or less than a third of the white standard, it is the one statistic that the American Black Community is rarely if ever fed from podia by rhetoricians striving to maintain their public visibility, who often resort to nonproductive self-serving theater.

Who has not heard at least once this comment?: "The black community earns $300 billion a year." This is more than the gross national product of the tenth largest country. It would be the ninth largest in the world. But, now that the American Black Community's present gross income has been overcome to where it would be only the thirteenth largest country, the black gross income statistic isn't recited much anymore. We'll discuss it for constructive purposes later on rather than for rhetorical garnishment.

Before The Speech—Message From The Grave

If Washington could visit with the American Black Community for just one hour, he would use that hour to teach us, as was his custom during his lifetime.

After getting our full attention, he would probably flash a reassuring smile we would regard as being from him to each of us individually. Then, he would lift his head slowly, almost imperceptibly without disengaging eye contact with us and might be heard to say barely audibly:

"For just a little while, give me the wisdom," as though drawing his strength from us. Then he would extend his smile and deliver upon us an infectious broad, toothy grin and begin his speech:

My dear Negro Americans, brothers and sisters.

Just on the basis of individual efforts alone, Negroes have made monumental gains in commerce and education. Monumental gains compared to Negro business statistics in 1900 when the National Business League was organized and in 1915 when the number of Negro businesses had tripled to about 157,000.

These are monumental gains compared to the number of blacks with college degrees in 1904, which was 2,500, and the number in 1942, which was 33,000. Monumental because today's figures of over 450,000 Negro businesses and over two million Negro college graduates are the end results of individual ambition. Tuskegee University, Bethune-Cookman College and Morehouse Medical School are the fruition of individual ambitions. Leland and other black colleges that folded reflect a lack of shared educational goals.

The National Business League was organized in 1900 to encourage Negro Americans to work in lock-step toward mutually agreed upon racial goals so that Negro growth could be monitored for both quantity and quality. The tripling in size of the Negro business establishment from 1900 to 1915 was not happenstance. It was an ambition of the league.

By 1920, the league and the Negro community were to have formed a partnership and become to a large extent mutually dependent. The partnership's effectiveness was proportionate with the understanding of the role each had to play and the extent to which it was played. The goals were to have been cultural and economic expansion proportionate with the Negro population and with every Negro committed to those goals.

Every Negro should know that with 450,000 Negro-owned businesses, the Negro community needs **over two million more businesses** to have the same proportionate share of businesses as white America.

Every Negro should know that with two million Negro college graduates, the Negro community **needs over five million college graduates** to be equitable with whites.

Every Negro should know that the Negro community's $350 billion yearly income must be **$565 billion—about $215 billion more**—to be on parity with white America's.

But, every Negro must be assured that these deficits can be erased, if erasing them becomes goals for the Negro community, and every Negro pledges commitment to attainment of those goals.

To inform every Negro of these deficits and to solicit every Negro's cooperation require total organization of the Negro community. The precondition for organization of the Negro community is an organized Negro business establishment, the Negro business establishment being the catalyst for qualitative Negro economic, educational and cultural progress.

American Negroes share no culturally endemic bond like the Jews whose common bond is Judaism, like the Arabs whose common bond is Islam and like most Europeans and most of those persons of European descent whose common bond is Christianity. So having none, one must be developed and made part of the Negro's daily fare, separate and independent of whatever culturally alien mores Negroes were forced to adopt.

I recommend a great national Negro monument in Washington designed and financed by Negroes. It would be the first national symbol and reference point ever to which Negroes can relate, none other of national proportion having ever been built with Negroes in mind. The monument would not only greet domestic and foreign visitors to Washington but would stand as mute testimony to the unheralded presence in America of unarguably its most indispensable racial group. An imaginatively designed monument could communicate to viewers evidence of that boast of the indispensability of Negroes in the making of America.

The catalyst for bringing the movement to full fruition should be the National Business League, not as one of its goals, but as its responsibility as the administrative center of the Negro Business establishment which it represents to the world.

Let me caution: To circumvent the league on this or any other project, where total organization of Negro America is necessary, would render inconsequential and futile any attempt by other organizations or persons to implement so massive a project and expect substantive results.

Since 1865, colored Americans have needed a bonding symbol. **Second to a bonding symbol, corporate empire builders are the Negro community's most crying need.** They are conceived and nurtured by the essential bonding symbol, endemic or artificial. The National Business League was to have been a producer of empire builders, but it lacked form, bulk and total Negro community involvement in its development.

The cross and Star of David are the bonding symbols of America's corporate empire builders. Their national reference points are the capitol buildings and executive mansions toward which they gravitate in times of crises and which historically have had little Negro relevancy, and are therefore not valid reference points for Negroes, again as history will show.

There exists among the bonding symbols, the empire builders and the reference points an interdependence or mutual reliance that can be understood and appreciated by others whose cultures are similarly stratified. Having no endemic religious bonding symbol, blacks must contrive a facsimile for racial ties with absolutely no severance of religious bonds.

Empire builders will emerge in proportion to

Before The Speech—Message From The Grave

Negro community support for a national black monument as a bonding symbol. Other national reference points will inevitably follow. **Now don't confuse ego builders with empire builders. Only empire builders are nuclei of power.** The time constraint precludes any further elaboration.

Finally, in sorting out and contemplating what I have been saying on this the eve of the 100th year of my 1895 Atlanta speech, without a doubt you will be aghast at the radical changes the Negro community must undergo in order to pursue the course I have chartered. But, you must pursue it. You got through the first 99 years of the severe American crucible with noble restraint before the most indispensable and pivotal person in American history, Martin Luther King, Jr., caused the playing field to be leveled.

Your pursuing the course I chartered will be the equivalent of Negro America putting a team on that field. Your present education, annual income and business profiles are the proverbial half loaf. The other half will require as a minimum total Negro involvement in a struggle not nearly as severe as the one King's generalship successfully ended.

So I counsel you: Pursue the course and reverse the Negro male's devolving into irrelevancy in the perception of Negro youth. Pursue the course and free Negro America of its reactive nature in favor of proaction, hence assertiveness. Pursue the course and risk the inevitable changes that will create the environment necessary to heighten in Negro youth self-esteem, self-confidence and ambition. Negro youth, because of their own distinctive nature, sing loudest, best, and with fewer inhibitions, following a loud and upbeat downbeat as would be inherent in pursuing the course I have chartered.

Otherwise, they succumb to a paralysis that mutes their voices and prevents their singing at all. The minority of self-motivated individual Negro achievers won't be denied, but the rest will be if they are not primed by facilitators. Pursue the course and thereby cast a warrior's shadow in which many will stand, many more will want to stand and all who do will regard it as a rare privilege.

Thank you, thank you.

I spent my middle childhood years, ages seven, eight and nine, and part of ten, in the small oil refinery town of Norco, Louisiana, 17 miles west of my hometown New Orleans. While there, I met an itinerant gambler, a Mr. Canoon from Alabama.

It was he who introduced the name Booker T. Washington into my life and hinted of Washington's historical significance. Mr. Canoon had attended Tuskegee Normal School in the early 1900's when Washington was its principal. He told me of his only personal contact with Washington. One weekend, he encountered "Professor Washington" while wandering aimlessly on campus and was asked why wasn't he ambling around downtown Tuskegee with the rest of the students.

"I told him the last time I was in downtown Tuskegee, I was verbally upbraided by a white grocer because I referred to Professor George Washington Carver as 'Mister Carver.' The grocer yelled at me that I had better never again refer to a nigger as 'Miss' or 'Mister' in the presence of white people. And, he said that included you also Professor Washington."

Before The Speech—Message From The Grave

According to Mr. Canoon, Washington was unruffled and simply smiled and said, "Then, son, when you are around them, refer to me as Booker as they do and to Professor Carver as Old George as they do. But, the problem is not what they do to us, but what collectively we allow them to do to us. You see, only a minority of us understand the meaning of racial dignity and racial esteem and how we are regarded as a people.

No Negro man today, no matter how influential and respected, can muster a significant number of Negroes to take a position in defense of Negro dignity and esteem. If you had persisted with the grocer in what you thought was your prerogative to indulge—referring to me and Professor Carver as 'Mister' or my wife as 'Missus,' a thousand Negroes would have looked on while twenty white men assaulted you, maybe even killed you."

Mr. Canoon then described how Professor Washington looked searchingly about the campus before he said almost in a whisper:

"Young man, if you see a white man enter the Negro community wearing a smile, he's coming to help you build a church, but if he comes without that smile, he's on a mission to destroy your businesses, maybe your community and perhaps even you."

Mr. Canoon was killed several days later "up the river," it was said, which meant any of several dozens of farming communities between Norco and Baton Rouge. He was stabbed, we heard, by another gambler who caught him cheating in a card game.

Occasionally, some non-white/non-black naturalized persons become genuinely offended at the heavy emphasis that is sometimes placed on black issues by the media. One such person recently reminded me there are other "minorities" in America besides blacks. But, he became visibly disturbed when I told him that most of those other non-whites recently arrived and owed Martin Luther King, Jr. a debt of gratitude for creating the climate that caused Congress in 1965 to liberalize the national immigration laws to allow more non-whites to emigrate to America.

"But," I said, "did those other non-whites arrive here as did blacks in 1619? Did they fight in the Revolutionary War as did over 5,000 black slaves and some free blacks, blacks without whom the colonial army of about 77,000 could not have won the war? Did they provide free labor for over 250 years to the mid-1860s building the foundation for what today is that great Edenic oasis of human freedom and incredible industrial complex called America?

Were they the object of the U. S. Supreme Court's shameful decision in 1857 in the Dred Scott case? Were they with Andrew Jackson in 1812 and Teddy Roosevelt in 1898? Were they at Argonne in 1918? Were they whom the Warren Court had in mind in its 1954 school desegregation decision? Or, were they whom Congress had in mind in passing the Civil Rights Act of 1964? Or, were they for whom the Voting Rights Act of 1965 was enacted?"

This person's response left me in a state of shock and raised my level of respect for him: "I am not very informed about blacks in American history or earlier U. S. immigration laws, and I apologize that my ignorance upset you. But, I truly regret that the degree of intellectual honesty necessary to infuse American whites with the moral courage and courtesy to celebrate American blacks for just those things you enumerated just does not exist."

I corrected him: "The whole world needs such an infusion."

"I can't disagree," he replied. "But, while I am not very informed about blacks in American history, is the average

Before The Speech—Message From The Grave

educated black informed enough to make the analysis you just did?"

My answer to him is not important, but what I have since wished I could have responded might be of interest to the reader. I wish I could have said that what I had just rattled off to him I had learned from reading Benjamin Quarles, John Hope Franklin, Lerone Bennett and E. Franklin Frazier, etc.

I wish I could have said that several months of the year have been designated by the National Business League for the commemoration on T-shirts of many of the historical events I recited as part of black America's economic and cultural growth initiative. I wish that I could have said 1880's newspaper editor T. Thomas Fortune is the cornerstone of black militancy in America.

In the 1880's, Fortune urged every black to own a Winchester rifle to use against anyone who threatened his security. And, in 1898 he commented that one white in Wilmington, North Carolina should have been killed for every one of the thirty-two blacks that whites massacred in their three-day reign of terror there to remove blacks from public offices, both elective and appointed. Blacks in Fortune's native Florida, and others elsewhere, have set aside each October to commemorate his October 3, 1856 birthdate.

June 25, 1941 immediately brings to every black's mind A. Philip Randolph and how he single-handedly caused that day to be an economic deliverance day of sorts for black America by literally having politically strong-armed President Franklin D. Roosevelt into issuing Executive Order 8802, which made illegal racially discriminatory hiring practices by military defense industries and the U. S. government agencies. Because of the generally shameful rendition of black history for domestic and world consumption, the reader will not find commemoratives of Randolph's momentous 1941 heroics that caused a significant increase in the black community's income, made college affordable to more black families and required more white businesses to accommodate black consumers.

However, two years later a beneficiary of Randolph's heroics, black Panamanian Hugh Mulzac, commissioned as the first black captain of an American liberty ship, the *S. S. Booker T. Washington*, was commemorated in a poem by Langston Hughes. History will show we are masters at trumpeting black "firsts."

Someone once said that the black community is so anxious to solve its problems, it forgets to first study them. I gleaned from the non-white immigrant's challenging questions that the American Black Community's dilemma is more serious than not studying its problems. It must know what its problems are before it can study them. And, after learning what they are, it must be committed to working toward solutions to them.

How will the American Black Community find out what are the problems? It will find out what its problems are through the emergence of black statesmen, a rare personality since Booker T. Washington. Leaders should emerge who speak directly to the American Black Community, tell it what problems they discern, suggest action blacks can collectively initiate to bring relief to and even erase some of them.

For instance, is the problem what we should call ourselves? Emma Lou Thornbrough writes in the book *T. Thomas Fortune, Militant Journalist* that in 1897 Fortune used the argument that Negroes, being African in origin and American by birth, should refer to themselves as "Afro-

Before The Speech—Message From The Grave

Americans." Blacks rejected that. It remained Fortune's term for identifying blacks until his death in 1928. But, in 1966 during one of the most defining moments in the American Black Community, the term "black" sprang from the freedom marchers in Mississippi.

Overnight it gained acceptance by Negroes in overwhelming proportions, and within a week virtually all black America had embraced it. The problem of what black Americans preferred as their racial designation was settled. "Black" was it. It now had historical connotations. It meant black America at its heroic and triumphal best, indeed in its finest hour. But, in 1987 or 1988, the term "African American" was offered to replace black, a gesture that meant its creators were sending a message to black Americans: "We know what's best for you, and 'black' is not. You're African Americans from now on whether you like it or not."

My non-white, non-black minority friend was furious: "What unbridled and shameless arrogance! How insulting!" But, I assured him that less than a majority of blacks use that term, even though the white print and broadcast media have allied themselves with those favoring its use—against the majority who have rejected it. What blacks wanted to call themselves was not a problem. A few highly visible trusted blacks made it one.

An occasion arose during the course of my conversation with the non-black/non-white immigrant when I could upbraid blacks like myself for allowing only a watered-down or D-minus interpretation of the measure of Martin Luther King, Jr. to commence filling history pages since long before his memory was commemorated with a holiday.

"If a history paper included his influence on the U. S. Immigration laws, it would rate better than a D-minus," the immigrant told me. I told him that when I was about five years old I asked my parents from where was my middle name Marcus taken.

"From Marcus Garvey," I was told.

"Who was Marcus Garvey?" I asked.

"He was a great colored man," my parents said, referring to Garvey in the past tense, although I was 19 years old when he died.

"Why was he great?" I naturally asked.

"Because he spoke up for colored people," James Oliver and Angele Bourgeois Pryor answered.

Now compare their reply to this current scenario nearly 70 years later of a nine-or-so year old black girl and a television reporter on Martin Luther King's Day:

> Reporter: What are you celebrating?
> Child: Martin Luther King's birthday.
> Reporter: Who was Martin Luther King?
> Child: He was a great black man.
> Reporter: Why was he great?
> Child: Because he spoke up for black people.

I paused for effect while the immigrant, I am sure, nervously searched for the appropriate response to assure me he understood the implications of the child's response. His rejoinder was classic: "That child is culturally fractured, seriously so. Who is responsible for that?"

I answered that his concern did not vary much from my comments here. That poor child and hundreds of thousands across America like her would not have improved on the image of black adults reflected in her answers had she alluded to the over-solicited "dream" of King.

The answer brings us back to the basic question, "whose responsibility is it to water our roots?" "If my parents, and logically my grandparents, were not valid root waterers, then it's obvious from our children's predictable responses that the problem has either never been recognized or has been avoided as too formidable a challenge. I believe the former since the premise of this book is that our perception of black history is faulty. We must, at all times, question the effectiveness of our teaching and the validity of what we teach.

King's legacy should be second nature to anyone who intends to articulate his greatness. Any person's history-worthiness is proportionate to his legacy. King's birthday was not made a holiday because he "spoke up for black people." Great minds saw his legacy as awesome and perceived its impact as boundless. But, it not being the nature of whites to exalt black heroes, except where it is financially rewarding, and even then cautiously, blacks must probe the depths of King's greatness and report them out: "mark one - mark two - mark three, etc."

King, in whose shadow all heroes of history stand and by whom all images pale, must have inspired Andrew Young's bold contention in an April, 1982 television interview that blacks have stimulated everything decent that has ever happened in America. His view had universal impact during his time and will impact more generations of the world than many not yet born will live to see. But, for Americans a simple statement of his legacy should be common fare in schools, church, at home and anywhere else, one that will be shorter than the Pledge of Allegiance and just as easy to remember.

Such a statement could be: An America without laws that favor one race, sex or gender over others. Or a slightly longer statement could be: Martin Luther King's legacy is an America free of racially restrictive and discriminatory laws and laws that discriminate on the basis of sex.

The previously mentioned immigrant agreed that blacks

undersell their history if my charges have credibility. Failing to plumb the full heroic depth and greatness of Booker T. Washington and A. Philip Randolph, a contemporary of King, became more understandable, he thought. It may have been more understandable but it was no less tragic and unforgivable, I reminded him.

The problem intensifies geometrically each King holiday celebration when his "I Have A Dream" speech is treated as the crowning jewel of his crusade. Evangelical in nature, it is at best superb theater: King's hopes and aspirations for all Americans sermonized with great passion to which the only reaction could be emotional. As a speech it did not approach statesmanship, and as a sermon, it was of little substance and hardly informative.

At the end of the "Great God Almighty, I'm free at last" part, the ages would have been best served had he motioned the crowd for silence and said in closing: "I've told you my dream, now you keep in mind yours. You too have a dream. Pursue that dream and work it through. Don't leave even one dream unworked to move on to another. And always before you have another to work, leave not a single dream behind you unworked."

The crown jewel of King's magnificent tiara of triumphs is the Civil Rights Act of 1964, which is the watershed of endless benefits to universal mankind for many generations to come. Besides America's changes in 1965 in the immigration laws already discussed, the great American market became accessible to foreign nationals outside of Europe, Australia and New Zealand.

The 1970's oil crisis enriched many Arab entrepreneurs who before 1964 were restricted to what they could invest in America. They commenced investing in Midwestern farm lands, and their buying frenzy gave the American business-political establishment the veritable jitters. The oil shortage forced America to open its small-cars market to allow more than Volvos, Fiats, Renaults, Volkswagens and other European-made cars.

The long-barred and scorned Japan-made cars were let in. It is ironic that Datsun's first national ad demeaned and insulted blacks: "Datsun, Dat's th' Un," or some such. **Japan doesn't publicly acknowledge its indebtedness to the American Black Community, nor has it ever acknowledged Dr. Martin Luther King, Jr., as the Japanese automobile industry's fairy godfather because black elders and erstwhile black scholars have not taught King as an economic facilitator of Japan or of other Asian countries.** Enough said?

No, not yet. There is an irony that must be related. The first time the American black people were referred to as a people by highly visible Japanese it was to insult them. In the late 1980s, Japanese Prime Minister Yasushiro Nakasone was reported by the media as saying blacks were among those "minorities" responsible for America's academic recession. He was followed by another politician, Michiko Watanabe, who was reported as saying blacks were lazy, dumb and deadbeats.

In 1990, black-bashing and denigrating were resumed with no less vigor by Justice Minister Serioku Kajiyama, whom the December, 1990 issue of *Black Enterprise* magazine quoted as likening blacks to Japanese prostitutes, who foul the atmosphere of residential neighborhoods. I have since asked many black owners of Japanese cars if these Japanese insults were taken into consideration when reparations were under consideration.

The Japanese must be told of Abraham Lincoln's, Frederick Douglass's, Representative Charles Sumner's and

Wealth Building Lessons of Booker T. Washington

Booker T. Washington's efforts in reparations for blacks, that is, compensation to blacks for services rendered by them in the making of America. Every now and then some lone black or several file reparations claims against the United States government. And, occasionally a white person or several will comment that they enslaved no one nor ever owned slaves, so why should they have to pay blacks for offenses they had no hand in?

The answer to that, of course, is quite simple: one can't be selective in what one inherits. Responsibility for America's multi-trillion dollar debt can't be disallowed by the yet unborn because they did not make it. One can't inherit only the wealth of one's ancestors. One is also heir to the ancestors' evils. Call that the law of social genetics.

But, the same logic carries over to reparation seekers: They can't be selective in whom they sue for reparations since slavery had universal sanction. Nearly all in the universe are defendants. Using Washington's 1895 speech in which he sought as reparations of sorts white businesses favoring the hiring of blacks over foreigners, I can construct a list of for what America owes blacks. But, the list of countries from which reparations would be sought would even shock the formidable and essential Booker T.

If intellectual honesty will be allowed to prevail over hypocrisy, then every country that was even marginally involved in the African slave trade would be named as a defendant in a reparations suit and appropriately billed as such. So in addition to America, most African sub-Saharan countries, most western European countries, some Arab nations and maybe even the Vatican would be named.

It becomes clearer then that reparations claims by one or several persons, or even by one or more organizations, would not be representative enough of black America to be effective. Nor can effective action be initiated by defensive leadership, which by its very nature is oriented to self-survival and whose gratuitous, unauthorized and unsupervised involvement would assure defeat of any efforts and set back the cause of reparations many years or even destroy all chances of it ever being pursued.

The only facilitating black-oriented body for initiating action in reparations action is the National Business League. Its involvement would mark the first time in the history of the 130-year-old American Black Community, the American black business establishment, 450,000 businesses strong, will be assuming its appropriate highly visible role as the bellwether in community leadership.

But, numbers representing potential power, the league's role must be orchestrated to attain desired if not maximum benefits. The power aspect must be developed and refined. Expressed for the American Black Community's benefit, its 35 milion blacks are its brigade of artillery. Its business establishment is its ammunition or firepower. No artillery piece is worth more than its metal without ammunition. This should better explain to blacks the genius of Washington in his emphasis on blacks building businesses and the commanding presence of the Japanese in the American marketplace.

America abounds with qualified persons, white and black, who would assist the league in developing plans to organize the American Black Community to expedite its move from a reactive body to a proactive one.

The move to proaction involves a number of activities: (1) soliciting readily identifiable black businesses to commit themselves to long-term support of the National Business League, (2) developing a plan for total organization of America's 450,000 black businesses, (3) developing plans for black America's national foundation for funding the myriad of activities that will be initiated as functions of its proactive mode, (4) developing a plan for organizing the American Black Community's 35 million blacks and soliciting their support for black America's national foundation, (5) conducting a national contest for design of the national black monument and a national administrative and communications center for black America, and (6) sometime in the future developing plans for a great cultural complex of museums, zoos, foreign pavilions and a family recreation complex and international sports center.

In a real sense, every American black will be afforded the opportunity to aid in the cultural growth of black youth and to contribute to the expansion of the essential black business establishment. The story of black America is not one of black criminality. Rather, **it is one of reacting to consistent external pressure from "white criminals" and their fellow travelers who relentlessly applied fierce resistance to the creativity of black people in so many areas, including business.**

But, interdiction of black entrepreneurial efforts did not dampen the courage and ambition of those caught up in the spirit of Washington's economic ambitions for black America. Thanks to Washington's legacy of wise counsel and the National Business League, for the first time in the history of the American Black Community, American blacks will be literally in the same struggle, knowing for what they fight and smug in the knowledge that the outcome will be **every black's victory or setback.**

Symbolically Booker T. Washington will have returned to live among us.

—TWO—

Booker T. Washington: An Uncommon Perspective[1]

Booker T. Washington's historical image has been seriously damaged. He has been labeled an Uncle Tom for instance, and his 1895 Atlanta speech has been called a "compromise" and a "sellout" among other things. Just about every recent treatment— written, film, oratorical— of his historical sojourn has been slanted to malign his memory.

Efforts now should be toward restoring to his name the dignity and esteem that once defined it and to get a national debate started on Washington to allow alternative perceptions of him to be introduced. To facilitate directing more black youths into rewarding pursuits, black history must be kept enshrouded in heroic garb. It is our duty.

We can't build on negative history. Building is a function of progress. In Washington-bashing, black history is being regressed into revisionist history, revisionist like

---

[1]Portions of this text was published in the Summer, 1993 issue of *The Boule Journal*, the official publication of Sigma Pi Phi Fraternity, Suite 703, 920 Broadway, New York, NY 10010. Reprinted with permission.

American history, which is being written as the miraculous product of white male ingenuity in total disregard of blacks' presence since America's founding.

Revisionist history is contrived, counter-cultural, anti-intellectual and designed to serve narrow and selfish interests and thus has no universal application. Washington-bashing is disrespectful of Washington's living descendants, the thousands of Tuskegee alumni and students who tapped into his incomparable legacy and the hundreds of thousands more who attended and are now attending schools named for him. Washington-bashing is irresponsible.

Ebony magazine's February, 1993 issue contained the results of a pool of 18 black historians and social scientists to determine who to them were the 50 most influential blacks in history. The results evidenced the effects of revisionism.

There is no intent here to convert Washington-phobes to Washington-philes so much as to present fresh and uncommon perceptions of Washington and his philosophy.

In 1865, the system of black slavery, incompatible with a free market economy, had been destroyed in the just-ended Issue of Black Slavery War—the name "Civil War" provides no hint of what the war was about, hence this name. But, destruction of the system of black slavery created another problem: A society in which all blacks were free was not compatible with the sensitivities of white Southerners who had operated the slave system and who were the losers in the war.

The majority opinion in the Dred Scott decision in 1857 held blacks to be an inferior class of human beings with no rights that white people were bound to respect. That opinion had become an article of faith for white Southerners and a basis for them to try to continue mastery over blacks.

Booker T. Washington: An Uncommon Perspective

Lincoln had succeeded in getting through Congress the Second Confiscation Act of 1862, under which blacks would be compensated with land as they became free, obviously for service rendered in the making of America.

Generals William T. Sherman and Rufus Saxon had already made awards of land to freed blacks under the act in their zones of operation when Lincoln was assassinated. But, Andrew Johnson, who succeeded Lincoln, voided the awards.

President Johnson opposed compensation or concessions of any kind to blacks, including the right to vote. He wanted them to just take their freedom and run. Not only did he order retrieval of lands awarded under the Second Confiscation Act, but he later caused the defeat of a resolution to require enforcement of the act.

Frederick Douglass was morally crushed by Johnson's actions. Douglass's great presence had obscured the presence of other well-known figures of the abolition era except the Harriet Beecher Stowes and the Abraham Lincolns. **I have linked Douglass here with Stowe and Lincoln as the second of America's three Indispensable Dominant Heroic Trios, the first having been Thomas Paine, Thomas Jefferson and James Madison and the third being A. Philip Randolph, Thurgood Marshall and Martin Luther King, Jr.**

Douglass's ulterior motive was for blacks to regain their inherent warrior instinct. Land was quantitative, tangible, and, given some, blacks would fight to retain possession. They had stoned federal troops who came to retrieve lands previously awarded to them. The Freedmen's Bureau and later martial law were mere bandages in a situation that required surgery, a euphemism for land ownership.

Hardly a year after the Issue of Black Slavery War, Southern white males initiated a genocidal campaign against blacks. In May, 1866, white males in Memphis sought out and shot whatever blacks came into view, wounded over a hundred and killed forty. Two months later in New Orleans, a mob of white males did the same thing, and forty more blacks were massacred.

A poem I wrote several decades ago sums up my feeling about the Freedmen's Bureau and the martial law eras:

The Debate

"In my mind," said Charley Cyrus,
"Progress is being made by us."
"Oh, off of that," screamed Harry Horus,
"Progress is being made for us,
Which to the short, 'tween and the tall
Is simply no progress at all."

Before the end of 1866, President Johnson had granted amnesty to rebels in such appalling numbers that any perception that the former Confederates had forfeited any constitutional privileges at all was illusory.

It was during President Johnson's administration on September 28, 1868 that the worst single carnage occurred. During the Opelousas Massacre 300 blacks were slaughtered. Few courts of law evidenced willingness to render politically "incorrect" judgements.

A mere five years after the war's end, all former Confederate states were back in the Union. Black votes gave Ulysses S. Grant the presidency in 1868 as they later did for Rutherford B. Hayes in 1876. Sensing a strategy to dilute black voting strength, the Republican-controlled Congress passed the Civil Rights Act of 1875. Opponents complained the bill was an encroachment on state jurisdiction.

In 1876 in Hamsburg, South Carolina, white militiamen stormed a cantonment of unarmed black militiamen and killed five. Two months later in nearby Aiken, white male terrorists massacred 40 blacks and wounded scores more. Then in 1877, President Hayes, mindful of the South's growing political influence, ended martial law there and also ended the policy of using federal troops at the polls. With those actions, the South's black community's die was cast. The fate of the Civil Rights Act of 1875 was a foregone conclusion. It was declared unconstitutional in 1883. The system was not working for blacks.

The overturning of the Civil Rights Act of 1875 was a pyrrhic victory for the South. It immediately commenced to pay dearly. It immediately began regressing into becoming the nation's cultural and educational cesspool, frequently adding to it putrescence black massacres and lynchings, crimes for which its white male perpetrators were rarely found guilty. And, where a guilty verdict was rendered, the punishment rarely fit the crime. Or, the offender was ordered by the court to appear at a later date for sentencing, orders that were mere formalities, not binding.

The ship that Thomas Paine, Thomas Jefferson and James Madison had launched and that Frederick Douglass, Harriet Beecher Stowe and Abraham Lincoln had set on a moral compass had been blown off course.

When the 1875 Civil Rights Act was vacated, Booker T. Washington had been principal of Tuskegee Normal School for two years. During a speech at the National Education Association convention at Madison, Wisconsin in 1884, he set the tone with the second item of his life's agenda for serving his race. He said, "At the bottom of education . . . of even religion itself. . . there must be for my race as for all races an economic foundation."

Then in 1886, the South received reinforcement of sorts for entrenching the Dred Scott syndrome. *Atlanta Constitution* editor Henry Grady, on tour to attract Northern industries to the South, addressed the New England Society of New York. In the audience were financier J. P. Morgan, the editor of the *New York Times*, statesman Elihu Root and businessman Charles Tiffany, among others.

Grady told the group that black and white schools in the South shared equally in education funds and that blacks received the fullest protection of the law there. Everyone who read or heard those claims knew they were at variance with reality, but that is what the people wanted said, no matter the validity. And, for the remaining three years of his life, Grady moved about the country saying the same thing and getting the same thunderous ovations and rave editorials as he had in New York. He softened attitudes toward the South by simply implying that the race issue in the South was being handled fairly and effectively.

In 1890, black editor T. Thomas Fortune organized the Afro-American National League to agitate for a better deal for blacks in America. The Henry Grady types were causing the non-South to become insensitive to the black plight and to slough over the race problems as non-issues. But, the league folded in 1893 for lack of black interest. It would be 19 years before Oswald Garrison Villard and some of his white liberal friends would form another league-type organization that everyone now knows as the National Association for the Advancement of Colored People or the NAACP.

The black bodies that carpeted the 30 years from 1865 to 1895 were not those of martyrs who died in defense of a cause to which they were irreversibly committed but blacks who were wantonly slaughtered by white males, not one of which was brought to trial and punished.

Black movie makers perhaps haven't the stomach to film such carnages as happened at Colfax, Coushatta, Opelousas and New Orleans on two separate occasions, which made Louisiana the black killing grounds of America's early genocidal era. The South as a whole assumed that distinction after Hamburg, Aiken, Mobile, Wilmington and bore it into the 20th century to add Atlanta, Tulsa and East St. Louis, among others, to its credit. Or, black movie makers haven't the skills required to capture the full heroic essence of those killed and the survivors.

A film on one or more such events might steer black youths off the too-often traveled fratricidal path. The film might cause black youths to modify their conduct in the war zones referred to here as Establishment-Induced Black Oppression Centers (EIBOC's) because the terms "inner city" and "ghetto" do not place blame for these economic morasses.

It was against this sordid historical backdrop that Washington was to speak September 18, 1895. That year was a landmark of sorts.

In addition to Washington's Cotton States Exposition Speech, Frederick Douglass died. Harriet Beecher Stowe would die the following year, and A. Philip Randolph turned six years old on April 15th. America's Indispensable Dominant Heroic Trios' accomplishments were **processes.** Black America's most momentous occasion, Washington's Cotton States Exposition Speech, was an **event.**

But as an elder, I must ask myself what **important lessons** should we derive from Washington's much herald statements? In addition, what experiences had Washington undergone that caused him to make these statements?

Wealth Building Lessons of Booker T. Washington

Therefore, using certain passages from Washington's speech (no doubt those that rankle some sensitivities in black America) I shall address the above two questions.

Washington said: One third of the South is of the Negro race. No enterprise seeking the material, civil or moral welfare of the South can disregard that many people and reach the highest success.
Comments: We're outnumbered two to one and hopelessly out-gunned as the thousands of blacks massacred by white males since the war stand in mute testimony. But, we're still here with you. If you move one notch, the able among us will move too, and we will pull some of the less able with us.
Washington said: Ignorant and inexperienced, it is not strange that when first freed we began our new life at the top instead of...the bottom. An elective office was more sought than real estate or industrial skills. Stump speaking had more attraction than starting a dairy farm.
Comments: He focuses building an economic base: real estate, dairy farm and extraordinary mental discipline. But, he chides his people for misdirecting priorities. Government was managed by plutocrats, he teaches. An elective office is not the office of the powerful but the facility through which the powerful deal. Black elected officials were without portfolios. They were Republicans out of gratitude, not because of political philosophy. And, they voted Republican out of gratitude, and thus they served the best interests of those to whom they were grateful. But, the bait of a black-occupied political office was too tempting to resist nibbling.
Washington said: Cast down your bucket where you are.
Comments: We [blacks] are experts in farming and animal

husbandry here. In the North, we will immediately become industrial illiterates. Our problems are here. Face them, and try to solve them. If we tuck tail and run from here, we'll be tail-tucking-runners-from problems there.

Washington said: We shall prosper in proportion as we learn to draw the line between the superficial and the substantial, the ornamental gewgaws of life and the useful.

Comments: There were black families he had visited in Alabama with $300 annual incomes living in lean-to, patched-up shacks in which were $1,800 organs but without a single blade to cut the grass that obscured their shack from the road. The danger was of such immature, irresponsible lifestyles being imitated by succeeding generations.

Washington said:. . . There is as much dignity in tilling a field as in writing a poem.

Comments: This is his triumphant gesture: he accorded recognition of the black community's largest economic class by affirming their intrinsic worth in one charitable and humanitarian overture. Of every 4,001 blacks, 4,000 had no college training. They were the laborers, domestic workers and field hands. They had provided the labor that built America. He told them, too, that education was not necessary for succeeding or starting a business. But, teach succeeding generations how to prepare themselves to attain lofty goals.

Washington said: To whites who look to foreigners for the prosperity of the South: You too "cast down your bucket where you are."

Comments: Here he jogs whites' sense of history. He said in essence that blacks were involved in all phases of the development of America. Then, he figuratively hands them an I.O.U. he had drawn up for their signature. He said blacks gave the work ethic meaning in building the roads and railroads, clearing the forests, tilling the fields that grew the foods whites ate, extracting ore from the mines, etc., all for free. Now that whites were paying for such services, they should not overlook employing blacks who they knew could do the work.

In addition, Washington was saying that the Negroes' presence in the Revolutionary War gave the colonial army the added strength it needed to win the war as by President Lincoln's own admission it did for the Union in the Issue of the Black Slavery War. He advised them to dispel any notion that if blacks prospered from employment, black men will despoil white women as white men despoiled black women.

Then he tried to allay white fears with his most misunderstood statement. He said: "In all things purely social we can be as separate as the fingers, yet one as the hand." However, the finger/hand analogy was obviously tongue-in-cheek for it was rendered invalid the minute a fist was formed or water was drawn from the well. The term "social" is too multifaceted to have definitive application.

Washington was a bit militant in the rebellious comment that in essence said eight million blacks will either aid you in pulling the South upward or make your pulling difficult if you disregard our presence. Several of Washington's statements require further clarification.

Washington said: The wisest among my race understand that the agitation of the question of social equality is extremist folly and that progress in the enjoyment of all privileges that will come to us must be the result of severe struggle rather than artificial forcing.

Comments: The actions of President Johnson dictated the terms for reconstituting the Union. A Southerner and a

Democrat, he blocked all black-enabling legislation or legislation that was offensive to Southerners' sensibilities. President Grant's loyalties had been to the Republican Party and the non-South's industrial establishment, one indistinguishable from the other.

The Republican Party needed numbers at the polls to offset the increasing voting power of the Democratic South, which Andrew Johnson had unleashed by the accelerated return into the Union of the former Confederate states. The Republicans also needed to develop a political and social climate conducive to industrial and economic growth and expansion.

Martial law was ostensibly to protect, entrench and extend Republican influence in the former Confederate states. Just as Democrat Johnson brought no white males to trial for the black massacres at Memphis, New Orleans, and Opelousas, so did Ulysses Grant's Republican administration avoid prosecuting white males for the many massacres of blacks at Hamburg, Aiken and other places that occurred under Republican leadership. Federal troops protected integrated Republican gatherings, and the military presence at the pools discouraged white violence against black voters.

Until history reveals that the death penalty or any penalty beyond the equivalent of a slap on the wrist was imposed on any white for crimes against blacks during the martial law period in the South, then the intent of the federal government was for reasons other than maintaining law and order. The integrated political gatherings were artificial and forced and given the prevailing societal climate would not have occurred routinely.

Except for Republican political strategy meetings, where blacks, carpetbaggers and scalawags of necessity had to function as an integrated body, it is reasonable to assume that martial law had nothing at all to do with maintaining law and order in the South in protecting blacks from white abuses.

It had more to do with insulating whites against criminal prosecution for crimes it was known whites would commit with reckless abandon against blacks and to assure the election and reelection of Republican candidates to state legislatures and Congress until the inevitable rise again of the Southern white male to assume command and directorship of the political process. Except for armed guards for integrated Republican political gatherings and assuring no interruption of the voting process at the polls, the federal troops had no real police function.

If the power barons of industry and the Republican Party were sincere about weaving blacks into America's political fabric on a permanent and meaningful basis, their strategy would not have resembled the short-term expedient that it was. We were being deceived and consummately had. We started out wrongly. Washington was admitting that we were not seeing clearly on the social issue. He felt it should not have been an issue at all.

The long lines at Freedmen's Bureau offices were figuratively being given bread and fish, not gardening tools and fishing gear. They were not being given firearms for hunting that were taken from the losers in the Issue of Black Slavery War. The losers brought them back home with them and frequently used them to hunt down and slaughter blacks who had been fringe victors in the war. But, Washington's prescription for fighting back was etched in granite: **build businesses, build businesses.**

Booker T. Washington: An Uncommon Perspective

Washington's actions were not rhetorical showmanship designed to elicit emotions and entertain. It was statesmanship that activates the creative juices. Marcus Garvey's creativity was activated, and he tried to construct an empire. And in 1968, it stirred the creative juices of Hartford's black business activists, and they proceeded to build a bank and organize the 300-plus black businesses that gave the local black community unprecedented productive economic leadership for more than three years, the most exciting years in Hartford's history.

White publishers would not publish Washington's Atlanta speech if it were given today. Why not? It is a blueprint for black upward mobility. If white America did not object to a progressive and upwardly mobile black community, there would be no EIBOC's.

Washington evidenced that day in 1895 a perception of America's sociological fabric held or admissible by only a few, black or white. Veterans of the Issue of Black Slavery War were as young as 42, young enough and in numbers great enough to have the edge in any additional campaign of terror against blacks besides the one then in progress. In the interest of the safety of his people, Washington could not risk a show of outrage or anger.

Any direct protests against the genocide then in progress would have been as effective as using eye-drops in a desert storm. The carnage was irrepressible because it had no opposition from law-enforcing establishments, local, state or federal or organized opposition from the Negro community. Even when the last war veteran had passed, an era of less intense violence would prevail as the residual effect of the racial venom sown by the Confederate veterans when they lived.

Washington's earlier warning that to deny blacks an equitable role in the South's striving would have deleterious effects was prophetic. He was anticipating the South's sordid circumstances after the inevitable decision by the U. S. Supreme Court in *Plessy vs. Ferguson*. But remember, the ship was launched by Paine-Jefferson-Madison and set on a moral compass by Douglass-Stowe-Lincoln. Then, it was blown off course by the 1883 ruling that voided the Civil Rights Act of 1875, and it commenced to assume the attitude of a rudderless pilotless vessel. It would be another 68 years, as Washington later predicted it would be, before America's Third Indispensable Dominant Heroic Trio would set it back on its previous course, sailing once more with a sense of moral purpose.

It is perhaps what occurred in Wilmington, N.C. on November 10, 1898 that caused Washington to formulate his next constructively strategic move. In a civil disorder there, erroneously called a race riot, 32 blacks were massacred by whites who also destroyed nearly twice as many black businesses and ordered blacks to get out of town.

Two years later, in 1900, Washington organized the National Negro Business League. **Today, as then, it is the most necessary, essential and <u>underutilized</u> national black organization.** Six years later the Atlanta riot, in which blacks fought back, confirmed his suspicion of what the real targets were during such disorders: **black businesses.**

At least ten blacks were killed, **but over 100 black businesses were destroyed**. Fifteen years later, in 1921, in Tulsa, Oklahoma, sixty blacks were killed in a race riot, **but over 100 black businesses were destroyed,** many by the police who bombarded businesses from planes they commandeered. From an estimated 30,000 to 57,000 black-owned businesses in 1900, the black business establishment grew to more than 100,000 businesses by 1915, the year Washington died. And, from about 7 million black-owned acres in 1900, the number exceeded 15 million in 1915 and commenced to diminish after that year.

Not only has there been no notice by blacks of whites' unrelenting interdiction of black efforts to build an economic foundation of some substance, but it has also gone unnoticed that in the 94 years since the National Business League was founded, a sitting U. S. president has never addressed one of its conventions.

Previously, I requested President William Clinton to consider addressing one of the National Business League's conventions. I stated that the purpose of his address need not entail announcing a federal black business program but simply to recognize the 450,000-member black business establishment, relate his office **sympathetically** to black America's efforts to build that all-essential but elusive substantial economic foundation, and thus become the first sitting U. S. president to speak directly to an American black business convention. The reply from the White House? The President had a full schedule.

On the other hand, ponder this: The father of the black self-help ethic is Booker T. Washington. Yet for over a dozen years, so-called "black summit conferences" on self-help have been held on occasions without one time invoking the memory of or extending gratitude to Washington and without the involvement of the National Business League, which was founded by Washington as the vehicle for facilitating black self-help. There has never been a book (until now) on self-help for the total black community that suggests how the league can be employed as facilitator of black economic advancement and be an extension of total organization of American blacks.

Washington would have disapproved black protests of Asiatics settling up businesses in EIBOC's where black businesses previously failed. He would counsel that black protestations should have begun, if they should have begun, with the movement into the EIBOC's of McDonalds, H.R. Block, Burger King, Kentucky Fried Chicken, Waffle House, etc. Washington would advise black religious activities and the eight black fraternities and sororities to capitalize respective stock corporations to accommodate some portion of black America's more than $300 billion annual income and to patronize blacks' desire for a respectable niche in the world of capitalism.

Washington worked from a well-constructed agenda from which he rarely digressed. In barely 20 minutes, he succeeded at Atlanta in articulating the most crying needs of his people and established at the same time a rock-solid basis for reparation claims at any time in the future.

In his "tilling a field/writing a poem" passage, he figuratively commingled the Big House bunch and field hands, mommies and mammies and "darling little dears" and pickannies into a conglomerate whole, which defined the American Black Community and coerced it to commence defining itself as a unity rather than "we" and "they."

Washington's suggestion of blacks' indispensability in the making of America was a bold stroke and his enumerating their contributions in that same bold stroke rendered him indispensable to any future plans for substantive progress by black America. Unlike his detractors then, and some yet to come, Washington did not speak from the relative safe haven of a big city or some northern perch. He spoke from the porch of the oppressors.

The challenge to black leadership now is to determine how Booker T. Washington intended to use the National Negro Business League to leverage the American Black Community into a respectable, substantial and commanding presence.

Booker T. Washington: An Uncommon Perspective

Often more can be gleaned from what was not said if it can be determined what the ulterior motive was in what was said. The league was surely intended to be multi-purposeful, and the benefits derivable for employing it for other than the obvious reason were meant to be infinitely more munificent. Failure to determine what Washington's extended plans were for the league would make it impossible to make any African connection under the newest fad "Afrocentricism."

For in retracing the black historical route through America, it would be impossible to get past the obstacle that has been created out of the Booker T. Washington segment of black history. With Afrocentricism now invalidated, would some other nonproductive pursuit replace it to detract from the more compelling challenge of organizing black America for forward and measurable progress? You provide your answer.

Booker T. Washington is the watershed of America's national black leadership and the creator of national black business leadership. Black Americans should demand that the black minority wrecking crew affirm these distinctions. Otherwise, they can't honestly and effectively state the black community's problems for those who must strive for a solution to them.

Black Americans should insist that black leaders and other highly positioned blacks get in step with the army of black tillers of the soil whom Washington ennobled in his 1895 Atlanta speech and who in turn immortalized him. Such mutuality cannot be ignored. It is strange that we who have not distinguished ourselves as builders are suddenly the leading candidates for the dubious distinction as wreckers.

Washington is the historical cornerstone of what will one day be Edifice Black America. All building

blocks that will go into erecting Edifice Black America of necessity must relate to the cornerstone, the most essential of stones. Be mindful of its importance: Washington's shadow is long, very long. Few qualify to stand in it, perhaps none of his detractors.

The next time we hear the chant, "Ain't gonna let nobody turn me 'roun'," consider this: Maybe it is best to be turned around since we might be going, if in fact we are going, in the wrong direction.

—THREE—

Questions and Answers: Part 1

After my article, "Booker T. Washington: An Uncommon Perspective" appeared in *The Boule Journal*, I received a large number of challenging questions and comments. They were asked over many months: some at conventions, some by phone, others by students during and following a lecture, by people who had either heard about or read the article and by people, friends included, whom I steered into a discussion about Washington.

These questions proved extremely valuable and my answers to them form the foundation upon which Chapters Three and Four are based. The questions are presented in the context that I can recall but not necessarily in the order they were asked, nor are these the only questions that were asked.

Question: You entitle your article, "Booker T. Washington: An Uncommon Perspective." It is my opinion that Washington's teachings are being widely accepted and translated into plans for action. Wouldn't that raise questions about your perspective of him being uncommon?

Answer: If what you say is true, then I am wrong about the results of the poll published in the January, 1993 issue of *Ebony*. And, I am also in error about an NAACP fundraising letter in which Washington's name was evoked with derision. What you say is at variance with what is happening in black America.

If Washington's teachings were being translated into action, so profound would be the changes in the black community that it would look like a cleanup crew had just been prepped to carry out Washington's legacy. The type of self-help he contemplated was not spelled out, it had to be developed using resources available within the black community, such as money, native ingenuity and functional intelligence.

The Washington legacy discourages bravado and the kind of emotionalism that led to the "do-your-own thing" craze of the late 1960's and early 1970's, which completed the disorientation of black youth and caused their nearly total loss of focus. If Washington's legacy were the black community's guideline today, the themes of clothing of adults and youths alike would broadcast it as a new sense of hope. "Back to Booker T!" would be black America's triumphant rallying call. The spirit of black America is somber now, pensive even and remorseful.

Question: What should black Americans do that is not being done?

Answer: Questioning black leadership. Without generalizing, how has black America benefited from your leadership? How is it better off? I think Booker T's ability to show the benefits his teachings had reaped for black Americans while he lived aroused the jealous streak in a minority of the scant few educated blacks then.

Questions and Answers: Part 1

I believe that had Washington in his Atlanta speech praised the 2,000 or so blacks with degrees in America at that time, he would have had almost no detractors while he lived. I think his slight of those proud achievers left a deep fresh wound to which the *Plessy vs. Ferguson* ruling eight months later figuratively added pepper.

Favored treatment status for educated blacks might not have lessened their fight for the uplift of the less fortunate of the black race but to be subordinated on the basis of race made a mockery of education as the cornerstone of an advanced society. Jealousy and rage combined to cause irrational action against the nearest and most vulnerable object—Washington himself.

That kind of black conduct was imitated by black youth later in domestic violence from the 1960's-1970's. With whom would Washington be replaced to manage the black business establishment's growth and nurture Tuskegee to college and university stature or graduate one more black youth trained in the industrial arts for the job market? And who of those at Watts, Hough, etc., would replace the hundreds of businesses destroyed and employ the hundreds of blacks who lost jobs? Who was irresponsible?

Question: You indict revisionist black history as damaging to American blacks, and you compare black revisionism to revisionist American history. But, America is doing quite well with its brand of revisionism. Isn't what's good for the goose is also good for the gander?

Answer: America's white community has 373 years of presence in America and a success-failure ratio that can allow for any history dilly dallying necessary, which will enhance its heroic posture, including ignoring blacks' presence or even lying about it. This means even fabricating events to make heroes of murderers, outlaws and the vilest of characters: bash a few here, build up a crud or two there, with thousands of characters to manipulate in relating America's history. No one can change what America has become or cares what roles whites ascribe to whom. There is no organized white predisposition to discredit substantive white players in the making of America, not even Confederate leaders of The Issue of Black Slavery War era. The American Black Community has existed for only 130 years. There have been only a few blacks of high visibility during those years whose legacies are definable, and Washington is one of them.

A person's legacy should determine his historical worthiness, and indeed, significance. There are too few blacks of substantive worth from black America's 130 years as a "free community." The whites have a surplus of expendables and can afford to trash hundreds, but among those trashed would not be a Benjamin Franklin, Thomas Jefferson, James Madison, Andrew Jackson, John Marshall, Abe Lincoln, etc. These people are white America's equivalent of the American Black Community's Booker T. Washington, Rosa Parks, A. Philip Randolph, Thurgood Marshall, Martin Luther King, Jr., and others.

But brace yourself: allegedly responsible, so-called black intellectuals have pummeled Washington's image into obscurity, ostensibly trying to chase him out of black history. Except for Frederick Douglass's contributions to Thaddeus Stevens's "Forty Acres and a Mule" for former slaves, in the late 1860s during Andrew Johnson's administration, the Great Liberator had no high visibility or pivotal role in the American Black community. So if there had been no Booker T. Washington during the post-Issue of Black Slavery War, the American Black Community history would be colorless, insipid and filled with minor characters who, with the exception of militant editor T. Thomas Fortune, A. Philip Randolph, Thurgood Marshall and Martin Luther King, Jr., rode the Washington Express into historical visibility.

Question: What did Washington have in mind for the National Business League? Specifically, what was to be its role?

Answer: The story of Hartford, Connecticut's Ebony Businessmen's League should be told in a book to best answer that. Less than 40 black businesses existed in Hartford's black community of about 35,000 when the league was organized in April, 1968. Its very existence was a magnet for hundreds of incipient business persons who saw in it a support facility, an image the local Chamber of Commerce could not project. **In the 15 years, the NBL existed under Washington's leadership, the number of black businesses in America tripled.**

Two years after Hartford's Ebony Business League was formed, the number of black businesses had increased nearly 1,000 percent. Collectively, as the Hartford-area black business establishment administrative agency, the Ebony Business League achieved the following:

1. arranged for business management instruction on an as-needed basis,

2. organized over 100 construction companies into an association to bid collectively on projects none could bid on individually and caused employment in the black aspect of that industry to more than double in several years,

3. sought financial aid for and directed the delivery of such aid to black businesses affected by two major civil disorders,

4. staged America's "first truly national black business show." This show was attended by nearly 60,000 people from May 7-9, 1970,

5. arranged business management classes as needed,

and assisted in preparing loan applications for all who requested the service,

6. caused a Junior Achievement company to be set up in the black community,

7. assured the publication of a local mostly-black high school's yearbook,

8. issued Thanksgiving food baskets to nearly 200 indigent families,

9. organized a women's auxiliary unit, which supervised the league's annual Christmas parties at community centers for as many as 6,000 indigent children.

These accomplishments and others translated into black business leadership, the first ever for a major American city in the 130-plus year history of the American Black Community. These were not spelled out by Washington. But, one reading of the wise counsel in his magnificent 1895 speech in Atlanta would suggest that black America must do whatever it takes to (1) build a respectable business base, and (2) direct the energies inherent in such a base into improving the quality of life in the American Black Community.

Question: We hear Washington told "darkie" jokes to prominent white audiences. Can that conduct be forgiven?

Answer: We're still laughing at Paul Laurence Dunbar, Flip Wilson, Redd Fox and Pigmeat Markham. Even now some well-known blacks tell race-demeaning jokes to whites and use the "n" word, which Washington seemed to have avoided. Paul Robeson is reported as having kissed his wife at a white friend's home. Noting some restraint on her part, he embraced her again and said, "Come on now, be all ("n") and give plenty." Many of us have had to flinch and even recoil because of racial jokes in the presence of whites. I thought we were being lured into the same mode of intolerance that defined white America before A. Philip Randolph's heroic stance in 1941, which caused the first crack in the wall of racial segregation, when I heard the negative reaction of some well-known blacks to other blacks who oppose affirmative action.

I was more than just aghast too at some erstwhile black scholars charging racism in the move to make participation in high school and college athletics contingent on meeting certain academic standards. To be concerned about "darkie" jokes by anyone in the American Black Community's formative years, but vote against functional literacy for black youth at this stage of the community's development, is to make a mockery of rationality.

That Washington stood on the dais with his fly unbuttoned will not be cause for him to be dealt with more harshly by history. But, to teach youth that academic literacy is inconsequential is cause for history to ignore the teachers. World-renowned black artists, whom I met at mixed soirees in Greenwich Village during the 1940's and 1950's, often waxed cute with "darkie" jokes. It happens even now.

While not regarded as chic, such conduct is still occasionally indulged. It has been, and is for some people we know well and least suspect, their bread and butter ticket. For many of our early, and even recent heroes, it was standard fare. Carter G. Woodson in *The Mis-Education of the Negro* indicted some mighty proper black folks for improprieties. Washington is perhaps guilty of such conduct. I might be too. But, I don't recall letting my contempt for such indulgence cause me to abdicate virtuous objectivity, make irrational judgements or cloud my reasoning.

Question: Of Washington advocating industrial education

for blacks—isn't that a downer? Wouldn't that tarnish his historical image?

Answer: It would if that position defined Washington's life pursuits. Perhaps the least heralded element of the Atlanta speech reads: "There is no defense or security for any of us except in the highest intelligence and development of all. If anywhere there are efforts to curtail the fullest growth of the Negro, let these efforts. . . be stimulating, encouraging and making him the most useful and intelligent citizen."

Unfortunately that jewel immediately follows the grossly overemphasized fingers-hand analogy. The New Orleans grade school system of my youth evidenced the Washington concept: one or two hours of industrial training under Mr. Norris each week at my school, the Joseph A. Craig Elementary School. The remaining hours of study were spent in history, music, arithmetic, English, literature and geography. The girls were given two hours of training in domestic science.

At another school, Albert Wicker Junior High School, Mr. Martinez and Ms. Edna St. Cyr conducted the two hours-a-week of manual or industrial arts and domestic arts. Those maintaining a grade of at least 85 percent for the first two years could attend McDonogh No. 35 Senior High School where laboratory physics, chemistry and biology were available, along with Latin, spherical trigonometry and solid geometry.

Otherwise, students took their third year of high school at the junior high where only French and Spanish were taught and no laboratory science was available. If one's grades were high enough, and one's economic circumstances ruled out attendance at a college, one could have applied for attendance at the normal school—tuition free—to be trained for two years for the teaching profession.

Questions and Answers: Part 1

Many of my friends dropped out after elementary school, some after junior high, for employment in the booming public housing construction industry, where Mr. Norris's and Mr. Martinez' training were applied. Some too, like myself, were hod carriers during the summer. For some, the Washington concept was a lift up the ladder to Tuskegee, then a full-fledged respectable college.

Mildred Balthazaar earned bachelor's and master's degrees in education there and embarked on a successful career in teaching. Norris Bucksell studied printing in the institution that Washington built, which culminated in a successful and highly rewarding career in the U. S. Government Printing Office.

Many did not take the road that led to Tuskegee, but from a Washington-influenced grade school curriculum they too had received a leg up the ladder to success and never lost their forward momentum. These included the late jazz legends Joe Newman, John "Picket" Brunis, Walter Pichon, Jr., Drs. Noel Gray, Joe Harding, Louis Blanchet, Jr., Leonard Burns and the late Joseph Boyer and Noella Pajeaud.

Washington held great respect for individual choice being a right not to be tampered with or denied. In May, 1900 at a meeting of the Bethel Historical Society, he took the position that full intellectual development of Negroes should not be restrained, for to do so would seriously impact the prosperity of the South. Is it possible that he anticipated world-class Tuskegee alumni like poet Claude McKay, writer Ralph Ellison, singer Lionel Ritchie, building tycoon Herman J. Russell and others?

When Washington founded Tuskegee, useful industrial trades were a logical first step to take in his estimation. As to his critics, who try to make his "darkie" jokes an issue this late in the game, he did what under the circumstances was necessary to keep Tuskegee functional and increase its viability in the interest of his people. Those critics should ponder this issue: Washington told "darkie" jokes in the interest of enhancing the quality of his race.

Many black entertainers today regress into "darkie roles and mannerisms" to enhance their personal fortunes. And finally, in 1942, of the estimated 33,000 black college graduates in the American Black community of 15 million or so people, what percentage of those were graduates of Tuskegee?

Question: Admittedly, much criticism of Washington might not have sound bases, but your supporting his position and seeming imperturbability about efforts by the South to legally deny blacks the right to vote is cause to wonder.

Answer: Such misconceptions disturb me. Washington seemed to have conceptualized his times. He weighed all facts bearing on the problems and decided to go for solutions or relief.

Question: Give an example.

Answer: In preparing the Cotton States Exposition Speech, Washington considered Mississippi having held a constitutional convention in 1890 mainly to deprive blacks of voting privileges. He knew too that earlier in the year of his epic speech South Carolina had rewritten its constitution and styled it after Mississippi's by basing eligibility to vote on land ownership, ability to read and interpret sections of the U. S. Constitution, having no criminal record, etc. But more importantly, Washington knew there had been no outpouring of black rage or ground swell of black demonstrations against such antiblack legislation.

For instance, there was no court testing of the constitutionality of the new provisions. And, Washington knew too that the reason there had been no black protest was

Questions and Answers: Part 1

because that part of Negro America that was in the South functioned as a police state policed by whites. Negroes were at grave risk who dared to exercise privileges, such as protesting, that whites took for granted. I am sure he reasoned as the Negro masses did: the time was not yet ready for such Negro militancy.

The National Afro-American League formed by T. Thomas Fortune in 1890 to showcase the determination of Negroes to secure their constitutional rights and privileges provided the best test of black readiness for asserting themselves in their collective interests. It folded after two years for lack of Negro interest and support. It was a paper army of generals with no troops.

The first time I read Washington's autobiography, I concluded that in the tradition of all great non-revolutionaries of history, Washington realized that he had not been commissioned by the Negro masses to speak for them. So, he first taught them, as the first third of his speech plainly shows by essentially saying it's a long way from where you are to where you want to be, but you can get there, if you do this . . . Then he pleaded with white America to give fellow Negroes, to whom all of America is indebted for its very existence, an equitable role in America's industrial expansion. No speech in American history equates the quality of statesmanship of Washington's Cotton States Exposition Speech.

Question: Did Washington ever publicly abdicate his indifference to the South's action to take the vote out of black hands?

Answer: Yes, but you won't read about it in many commercially oriented black-authored accounts of black history. In 1898, Washington anticipated the purpose of the Louisiana Constitution Convention and wrote a letter to the convention, which received press coverage. He pleaded that the voting law be fair and that it not be interpreted to favor the white man on the one hand and disfavor the Negro on the other.

There is no future for a state, he said in essence, which has a large percentage or segment of its citizens living in ignorance and poverty with no interest in government. Of course, the infamous "grandfather clause" did exactly what he asked not be done. It excluded Negroes because they were sons and grandsons of males who were not entitled to vote on January 1, 1867, more than three years before ratification of the Fifteenth Amendment that forbade denying any citizen the right to vote.

Then in 1900 in a speech in Washington, D.C., Washington said that while he advocated industrial education as a necessary first step for Negroes, full intellectual development of his people should not be artificially interdicted. When Negroes' right to vote was under attack, he felt compelled to speak. To deny any man the right to vote amounted to subversion of democratic principles and to permit ignorant whites to vote while denying ignorant Negroes that same privilege was unfair to both Negroes and whites.

And little is rarely said of Washington's behind-the-scenes activities to test the Louisiana suffrage laws where he raised what money he could from organizations and friends and made up the difference from his personal finances. The lawyers engaged to handle the case failed to prepare the briefs on time, so the case never reached the court. He also anonymously paid the total cost for contesting Alabama's voting laws in court, a magnanimous and unprecedented commitment by a high visible black in the interest of his people. In a majority decision, the U. S.

Questions and Answers: Part 1

Supreme Court ruled that the issue was a political one and therefore had to be corrected by the state legislature.

Question: How do you explain Washington's repeated efforts to keep his association with an activity or organization under cover like his association with the suits regarding Louisiana and Mississippi voting laws and ownership in certain newspapers of his time?

Answer: It should not be forgotten that as founder and principal of Tuskegee Normal School, Washington's overall success had drawn national attention. He seemed to have been resisting efforts to force him into a role he had not voluntarily chosen. He complained over and again that he was not a public agitator for Negroes' rights, which it would have appeared he was if he had exposed or admitted to his numerous involvements. He savored the recognition he received from the Establishment. Most admit that he deported himself commendably.

But, from all I can glean from what I have read about him, he was a leader from an education standpoint to the extent that he was a normal school principal. He was a business leader to the extent that he was president of the National Negro Business League and had the consent of America's Negro business establishment to speak for its members. His relationship to the American Black Community was as a teacher, which he evidenced quite convincingly in 1895 at Atlanta. It was Washington, and no one else, who said he would teach anyone who wanted to be taught all that he knew.

What stress he must have undergone to be sought out for his counsel on every imaginable issue by Presidents, corporate heads and foundation directors. They sought not just his counsel but his approval on actions they were planning. Imagine what went through his mind each time he fulfilled a request, wondering what would be the public reaction if from his lips some inevitable grievous human errors had fallen. What stress he also must have undergone because of the literary barbs and editorial snipping of the black press, which failed its readership by not sorting out for them who the real Washington was. Nothing in all I have read of Washington describes him as a willing leader of the Negro masses. I recall a similar circumstance in my life as the leader and spokesman for the Hartford 300-plus-member black business establishment from 1968-1972. Through no personal design or self-serving ambition, my advice, counsel and often consent were being sought on matters and about issues with which I was not conversant. It was an appearance here, a speech there, a meeting with city or state officials somewhere, etc. Some comments to the press were scrutinized and understood out of context. There was all of this because I had developed a level of leadership as new and alien to black Hartford as it was to all of black America. It was black business leadership.

Like Washington, my high local visibility was construed within the parameters most familiar to observers. I was a local black leader and therefore a *civil rights* leader. To my political party, I was a ripe candidate for public office (and I did run), to governments I was ideal for committees and commissions (and I was appointed) and to the private sector—community organizations, chamber of commerce, hospitals, etc.—I was of use to their purposes (and was used).

Even the school board asked me to sit in on a case involving a tenured teacher and a teacher's aide. As a result, my assertive leadership mode was modified for less productive and rewarding societal demands. In time, my identity as a business leader gave way to community leader. The die was really cast when I was appointed by the governor to the state's human rights and opportunity commission.

Though Washington successfully resisted services that by their very nature would have earned him designation as national Negro leader, except for accommodating his people with an occasional speech before odd and sundry black organizations and activities, he is still regarded as an educator and national leader. His measure is determined on those bases and rarely as a sage and elder of the American Negro community, a black educator and a national black business leader.

Until black history deals with him in those contexts, then it is presenting readers a contrivance, a phantom or a money-maker that rival Davey Crockett and Daniel Boone in commercial appeal. That's some of the revisionism that caused me to present the article. Black America deserves better, as does the world.

Question: Was Washington not speaking as a Negro leader in his Atlanta speech when he said "16 million Negro hands will help the South in its growth efforts, or the same 16 million hands will pull against the South and retard its growth?"

Answer: Taken literally, he is. As a waterer of black roots, that is as a black elder whose responsibility is to teach black youths, I would fail in discharging that responsibility to give a literal interpretation of that statement. He was both the prophet and teacher here, prophesying the fate of South if the Negroes' share in the South's industrial growth were not equitable in this poem:

> The laws of changeless justice bind

Oppressor and oppressed;
And close as sin and suffering joined
We march to fate abreast.

The lesson to blacks was basic: the prophesy was an hypothesis and would not simply happen. It had to be made to happen. A precondition for making it happen was total organization of black Americans. Without total organization, ten million Negro hands could not be raised, pressed or applied in any way against anything with achieving predicted results in mind. Whatever ills the South would undergo for disallowing Negroes a respectable share of its successes would be passive, disproportionately less than the societal ills such slight would cause to be visited upon Negroes unless they organized.

It was near the end of the 19th century, 1896, that Washington, seeing total organization as not forthcoming, prophesied almost correctly that it would be 50 years before the racist attitude of white America toward Negroes would commence changing for the better. In that he again was teacher-turned-seer, but that did not make him a national Negro leader. Nothing indicates that he ever was.

Question: Why didn't Washington attempt to organize the black community?

Answer: There are quite a few conditions that must be met before that can be done. I think Washington began working in that direction when he founded the National Negro Business League. I think he didn't really know what those conditions were, but he knew that somehow Negro businesses were part of the solution. That is a problem we must solve.

Question: We know that he never left us a clue as to how the black businesses figured in. Have you figured it out?

Answer: He left not just a clue but his reputation on black-owned businesses being the panacea for black ills in a 1894 speech in Madison, Wisconsin to the National Education Association Convention. There Washington said essentially that for the Negro race, as for every race, there must be an economic foundation, perhaps even ahead of religion. In other words, no priority precedes an economic foundation. From that lesson alone, I developed the following *bon mot*: "The name of the game is trade and commerce; no trade and commerce, no library, no city hall, no church, no school, no town—nothing." But, I also learned this lesson: From an economic foundation, all strategies for growth and development emanate. Otherwise, there is no autonomous growth and development.

It is imperative that Washington's Atlanta Cotton States Exposition Speech be recalled again. In that speech, Washington **hammered home the significance of a business base.** Recall he said, "Let down your bucket in agriculture, mechanics, domestic services, commerce, the professions. . ." **The full spectrum of what gives a community upward mobility is commerce**, and keep true the focus he had revealed at Madison. Acquire land, he taught. Learn the difference between what is superficial and what is substantial, between what is absolute junk and what is useful. Don't run the risk of opportunities passing you by while you are feeling sorry for yourself. His autobiography, *Up from Slavery*, should be standard fare in all black households.

Question: Washington prophesied near the 1900's that race relations would commence improving in 50 years. That would have occurred in the middle to late 1940s. You said he was right on target. Give some evidence.

Answer: It depends on what you want to use as a sign of

a softening of race relations. Consider these examples: (1) Randolph's daring facing down of President Roosevelt and causing him to issue Executive Order 8802 on June 25, 1941, which put an end to discriminatory hiring practices in defense industries and federal agencies, or (2) President Harry Truman's ordering desegregation of the Armed Forces in 1948.

The next question should be why did Washington arbitrarily select 50 years? Well, there was nothing arbitrary about his selection. He was not an emotionalist but contemplative and insightful. He factored in the Supreme Court ruling in *Plessy vs Ferguson*, the Dred Scott and 1883 civil rights decisions. It had not been unanimous. The minority opinion by Justice John M. Harlan, a Southerner no less, was a strong statement that bared the reality of American society versus constitutional proclamations and would influence future justices to break ranks with the unholy mindset of the Oliver Wendell Holmeses and Roger B. Taneys in whom the white-is-might-is-right syndrome was inextricably harbored.

Washington factored in too that the American Black Community was not a hotbed of agitation for legal rights and privileges and saw no signs it intended to deport itself. Remember, he made this prophesy just nine months after the Atlanta speech, which had raised him to national prominence and a month after *Plessy vs. Ferguson*. Like the speech, the prophesy was made against a knowledge of the American Black Community's history. Before the speech, his was not a household name.

T. Thomas Fortune, editor of *The Age* newspaper and organizer of the failed Afro-American League, was better known. It was the folding of the league for lack of black interest and black financial support, despite black civil rights

Questions and Answers: Part 1

and privileges being snipped at after the Civil rights Act of 1875 had been voided by the High Court, that would logically account for Washington's factoring in black non-assertiveness to make his prophesy.

Washington was a new voice on the national scene, and any notion that he was powerful enough to influence the Supreme Court so early in his era defies rationality. One friend playfully nudged me on hearing that: "Don't try confusing me with facts. My mind is made up." Washington's intercession with white America, on behalf of the overwhelming black majority, and the strategy for succeeding, indeed surviving, that he suggested for that black majority, were the only alternatives available to him. Without orchestrated black assertiveness, breaches in the wall of racial segregation would for the most part be by attrition only.

Question: Explain why any strategy for black growth and development must be against the backdrop of totally organized black businesses?

Answer: One of the profound statements that defines a capitalistic society was made in Washington's first speech before a large audience. It was in 1884 before the National Education Association Convention in Madison, Wisconsin which I referred to earlier. I read it in Samuel R. Spencer's book, *Booker T. Washington and the Negro's Place in American Life*. Washington said brains, property and character would gain for Negroes more respect than laws, but economic progress was the real harbinger of power: "At the bottom of education, at the bottom politics, even at the bottom of religion itself, there must be for our race, as for all races, an economic foundation, economic prosperity, economic independence."

From that statement, I developed one which I recite at

every propitious opportunity: "Every community is not all business, but every community is because of business." There is in every business a great sustaining energy. Efficiently harnessed and appropriated, the energy in over 450,000 black businesses is a great resource available to the American Black Community for possible use in any ambitious racial project. But, it is useless to our efforts in its present fragmented state. That is why I insisted on using the term "black business establishment." That means 450,000-plus black businesses.

When I read Washington's statement in Spencer's book that "We ask help for nothing we can do for ourselves," the concept of mass dynamics as a strategy to bring relief to many of the American Black Community's problems came to mind. That required reaching the total black population, with all having the opportunity to commit their efforts to attain racial goals. One friend said the church was the way to all blacks. I corrected him: all blacks did not go to church, but all blacks went to businesses. So, first the businesses must be organized, and then organization of 35 million blacks will be relatively simple.

Question: You seem consumed with "total organization." Why is that so important?

Answer: The black community is broken and needs fixing. Fixing it cannot be accomplished by one organization or even several, nor by one person.

Question: Hold on, now. You have been emphasizing the National Business League role as though it is the one organization that can do the fixing that needs to be done. What am I hearing now?

Answer: The National Business League would be that organization to which all black businesses relate and would be figuratively the administrative center of the black business establishment. It would be the facility where the American Black Community and its business establishment connect. Except for initiating the project designed to put the American Black Community in an assertive posture, the league's role will be to secure the participation of the majority businesses in black projects and keeping the black business establishment relevant to the American Black Community. But in its initial role, it would demonstrate to black Americans the rewards that can be gained through total organization.

For instance, black Americans have no concept of a national black administrative complex located in Washington, D.C. This would be symbolic of a black America in a new and assertive role: proactive at least rather than the traditional reactive role. Nor has black America previously envisioned a national black monument built and financed by blacks to be their symbolic welcome to foreign and domestic visitors to Washington, D.C.

The monument would have an American Black Community population counter in plain view to show from how many the welcome is extended. There would be a building in the complex that houses the American Black Community Foundation, among other things. Blacks who for so long never had much are not liberal givers to worthy causes. Many colleges closed or nearly did for lack of black support. Martin Luther King, Jr., for all his great works, often had funding troubles in his crusade that would never have occurred if blacks had given him representative or even moderate financial support.

The black business establishment should have been the catalyst in loosening the purse strings of American blacks to give substantially to an activity that was unquestionably meeting with great success. With the black business establishment involved, the net effect would have been King's efforts evolving into a national black civil rights movement rather than the non-black funded regional or series of local activities that it really was. One percent of the American Black Community's annual income then would have been over $400 million. King meant more than $400 million to any group that feasted at the trough of illimitable benefits to blacks, white women, non-black minorities, etc.

The black business establishment would be the catalyst in loosening black purse strings to facilitate organizing black America. Maybe this would not be to the extent of one percent of its gross annual income, although a larger percent should be the goal. The total gross annual receipts of America's 450,000-member black business establishment is $25-30 billion. From that comparatively small source necessarily will come the seed money—millions of dollars—for initiating the process of organizing blacks. Why go to a $30 billion source rather than a $300 billion source?

In a newsletter I once published, I wrote, "The extent to which a percent of the money spent in the community is returned back to the community is the extent to which that community is culturally and economically healthy." The majority business establishment validates that hypothesis through support of Lions, Rotarys, Kiwanis, Boy and Girl Scouts, civil rights organizations, foundations and cultural activities and in-kind contributions like the loan of employees to worthy activities such as Junior Achievement, etc. Often, the local Chamber of Commerce is the initial point of contact with the local corporate establishment for community activities.

The National Business League would be the initial point of contact with the American black business establishment for organizing that establishment and ultimately black

America. At the league, uplift efforts for blacks are at long last in the proper channel, ensconced in fertile grounds where growth and progress define all efforts.

The country's top progressive thinkers, and "movers and shakers," can be summoned by the league for planning, developing and directing uplift strategies. A suggested schedule of contributions for the 450,000 black businesses would be $15,000 for businesses with gross annual receipts of over $1 billion, $10,000 for between $200-999 million, $100 for gross receipts under $50,000 etc. There would be a timetable for inviting America's majority businesses to participate as would be a plan for explaining the objectives of uplift and soliciting the participation of blacks and "friends of the race."

It was with just such a progression of events in mind, inspired by my first reading of *Up From Slavery*, that I developed in 1968 the following slogan for Hartford's Black Business Establishment: "Taking the Lead, Setting the Pace." I hope that sufficiently answers your questions.

Question: From what I can figure out, some national organizations would become casualties if such a scheme actually materializes.

Answer: That is not the purpose for total organization of black America. All national black organizations would share in the contributions made by an organized black America. **In fact, by receiving funds, such organizations will in effect at long last be mandated to speak for black America, an authority that has never been conferred on any high profile civil rights advocate. So from the outset, a secondary objective of the uplift program will be to enhance all black organizations and their leadership.**

Question: Washington is supposed to have opposed or-

ganization of the NAACP. If that is a fact, why did he?
Answer: Black America's previous attempts to form national black advocacy organizations had not been favorably received. In Nashville, in 1887, the first notion for a national organization did not get beyond the talking stage. In 1889, T. Thomas Fortune formed the National Afro-American League, which Washington had shown excitement for when the idea was first advanced two years earlier. The league folded in 1892 for lack of interest by blacks. But in 1898, at the request of many well-known blacks, it was reorganized under the name National Afro-American Council. A year later a group in Boston who opposed Washington's views on the problems the Negro faced, organized another organization called the National Colored League.

Pro-Washington people—and Washington himself—caused the premature death rattle of the body. The council lasted for eight years and held its last convention in October, 1906. It folded shortly thereafter.

What distinguished the council and contemporary competing black organizations? They lacked the support of the black masses. Washington, considered a spokesman for the black community, had sensed that some of the most vocal of his critics wanted him to agitate for Negro rights. He had the support in the council to steer proceedings away from partisan politics, support that the intellectuals could not attract to their side.

At a meeting of Negro New Englanders, William Monroe Trotter, a recent Harvard graduate, opposed discussing business at all in favor of devoting all sessions to discussing the political rights Negroes were being denied. Only when the Negro has all rights granted him in the Constitution should business be discussed. He became

Questions and Answers: Part 1

Washington's severest critic. If the National Afro American Council had been in essence rejected by Negro Americans as evidenced by their failure to give it their financial support, then the NAACP, in process of being organized to agitate for Negro rights without soliciting their approval, was without validation.

Washington favored organization of the National Urban League because it was not political. It can be seen that his philosophy for advancement of blacks was intractable, and he thought it was easier to delay the approach of dawn than to stay the injustices Negroes were being subjected to, not just in the South, but nationwide.

His counsel to his people was to build businesses, learn a trade, and make themselves indispensable to the community in which they lived. To that end I think he felt that any time spent in criticizing him was time and energy wasted, energy that could have been more efficiently expended. Every generation of blacks since can testify that in the early years of the American Black Community, of the very few educated blacks there were, almost all were seriously deficient in prophetic vision.

The questions NAACP historians must answer without generalizing are: Did the organization facilitate whatever alleged gains black America made since its inception? Or, were the gains attritions, gifts of time, that is, bound to have happened had black America simply sat and waited? Was the NAACP a wet nurse to a situation that required a surgeon? Could it have earned surgeon credentials if its white founders had prior consent from Negro America to form the organization? How many more Negro voters resulted because the grandfather clause was ruled illegal? How much more integration of neighborhood occurred after removal of restrictive covenants?

If those questions can be dealt with and the answers to them analyzed, it might be found that Washington's opposition to just another Afro-American Council-type organization had considerable merit.

Question: About Washington prophesying the easing of racial restrictions, doesn't that require more than just your say-so?

Answer: In June, 1896, a month after the Supreme Court's ruling in *Plessy vs Ferguson*, Harvard bestowed an honorary Master of Arts degree on Washington. In his acceptance speech, he said in essence: The American Negro race must measure itself by American standards by which measure it rises or falls. Sentiments don't count. During the next half century or more, Negroes must pass through the severe American crucible to be tested. They must be tested in patience, forbearance, perseverance, their power to endure wrong, to comport themselves in the face of temptation, to economize and acquire skills, to compete and succeed in business and to disregard the superficial for the real, the appearance for the substance. . .to be learned, etc.

—FOUR—

Questions and Answers: Part 2

There were many questions and some discussions about William E. B. DuBois among the Booker T. Washington-related questions in Chapter Three. Black Americans have been so conditioned that a mention of the name of one triggers recollection of the other. DuBois was a contemporary of Washington, manifestly brilliant and well educated. He was the first American black to earn a Doctor of Philosophy degree from Harvard University.

DuBois gained high visibility during the early 20th century by writing a semi-autobiographical volume of essays entitled *The Souls of Black Folk.* In one essay, "On Mr. Booker T. Washington And Others," he took issue with Washington's advice to black Americans to forego political action and concentrate on learning a trade, accumulating land and wealth, and building businesses.

Question: You seem to have completely ignored DuBois in your article. Was that intentional?

Answer: One does not mix apples with oranges. Washington's legacy is numerous: a school, the National

Business League and a blueprint for building the American Black Community into a competitive entity. DuBois's legacy is a body of information mostly about himself and his times.

Question: How do you see DuBois?

Answer: He had an opinion about anything and everything but a plan for nothing. But then, why should he have had a plan? He was a chronicler and wrote mostly about incidents and issues affecting blacks of his time. Where in black history does DuBois fit? One's history-worthiness can be determined from one's legacy. What is his legacy? Since he was an educator, is it a body of knowledge or a philosophy of education?

I have read hundreds of his articles and essays, and they all share several things in common: unrelatedness, disjointedness and an element of spontaneity during which his brilliance at recall is brought into play. The American past and the American present are replete with DuBois-types: Walter Lipmann, T. Thomas Fortune, Henry Grady, Joseph Alsop, George S. Schuyler, Carl Rowan, George Will, Westbrook Pegler, and others.

DuBois's 1935 *Negro Reconstruction* work is not a widely used reference text which adds a new and significant perspective to the intrinsic worth and indispensability of blacks in the making of America. America would have ranked him with the immortals had he been a white man who wrote that. But, he wasn't. In fact, not many people even quote DuBois today.

Question: I do. Didn't he say the problem of the 20th century is the problem of the color line?

Answer: He published that in 1903 in *The Souls of Black Folk*. Emma Lou Thornbrough notes in her book, *T. Thomas Fortune, Militant Journalist*, that in 1897 Fortune was crusading to have "Negro" and "colored" dropped as racial

Questions and Answers: Part 2

designations in favor of Afro-American. Failing in his crusade, he wrote an article, "The Latest Color Line," from which Thornbrough extracts these words: "No friend of the Afro-American race can fail to regret that the black and yellow people will have the problem of manhood further complicated by a color line."

That was seven years before DuBois's magnanimous prophesy. But, let me relate something Thornbrough pointed out that is, in fact, attributable to DuBois. DuBois made a comment at a meeting in Boston expressing his opposition to the federal election bill that would have penalized states for not conducting fair elections. It was folly to think that laws could set right any wrong. He said:

"We must ever keep before us the fact that the South has some excuse for its present attitude. We must remember that a good many of our people south of the Mason and Dixon line are not fit for the responsibility of a republican government. When you have the right sort of black voters, you will need no election laws. The battle of my people in the South must be a moral one, not a legal or physical one."

We might pause before castigating him for the election bill statement since he was just a 22-year-old student at Harvard then, but he was 35 when he published the outdated "color line" statement and knew that Biblical Noah established the color line when he cursed his son Ham and said all of his descendants would be black. DuBois knew the color line was used to determine who would be human chattel for the New World slave markets.

He knew that Chief Justice Roger B. Taney drew the color line in 1857 in the Dred Scott decision that lasted until 1964. He knew that the courts ruling in 1883 on the Civil Rights Act of 1875 and in *Plessy vs Ferguson* in 1896 figuratively gave the color line a coat of fresh white luminous paint.

Before DuBois was born, the battalion of free mulatto Negroes from Louisiana, which fought for the Confederacy, evidenced awareness of the color line. So what was the purpose of DuBois's alerting his readers to that imminent timeless social plague, and what logical reason can anyone give for using that non-prophesy to enhance his historical image?

Question: Wouldn't DuBois's having founded the National Association for the Advancement of Colored People (NAACP) be sufficient basis for high visibility in black history?

Answer: If indeed he founded the NAACP. He didn't. There are several stories about how the NAACP came about. But, the most recent is told in David Levering Lewis's book, *W.E.B. DuBois: Biography of a Race, 1868-1919:* "The widely held perception that what became the NAACP was started by African-Americans is understandable but only symbolically true. For that very reason, DuBois deliberately inflated the contributions of African-Americans when he chronicled the events of the 1909 National Negro Conference. But, Ovington and the millionaire Socialist Walling, rather than DuBois and Trotter, were the sparks of the association, and Charles Edward Russell and Oswall Villard were the engines."

My recollection of what brought about the idea of what ended up as the NAACP is not one that is widely known. The story I recall is that Oswald Garrison Villard was thoroughly disgusted with the antics of DuBois and the group he pulled together at Niagara. His comments in essence were what they are doing might be all right for them, but it is not all right for a Garrison. He forthrightly wrote letters

Questions and Answers: Part 2

to some of his white liberal friends, believers in democracy, and a few prominent Negroes, DuBois among them, for a national conference in New York on Lincoln's 100th birthdate to renew the struggle for civil and political liberty for all.

Forty-three people attended, of which six, DuBois included, were black. Three months later in May, the NAACP was formed. There was just one Negro incorporator, DuBois, and only one Negro officeholder, DuBois, who was named director of publications and research. In that capacity, he developed *Crisis* magazine and managed it into a respectable and qualitative monthly Negro publication and it thrusted him into high national visibility. From that scenario, can anyone be coerced into believing that DuBois as the founder would have entered into any compromise that would have resulted in an all-white administrative staff or only one black incorporator?

Question: Wasn't the NAACP modeled on the order of the Niagara Movement? Doesn't that count for something?

Answer: That just might be another of DuBois's deliberately inflated accounts of his contributions. The first civil rights organization, the Afro-American League, was a model for the second, the National Afro-American Council, which was a model for the third, the DuBois Niagara Group, which was a model for the fourth, the NAACP.

Question: Why is so little ever said of the National Afro-American Council? It seems pivotal in our early history.

Answer: Just four years before the *Souls of Black Folks* was published, DuBois publicly defended Washington in a manner akin to worshipping him at the 1899 National Afro-American Council Convention in Chicago. A resolution condemning Washington had failed to pass and another honoring him had been offered and was adopted, ostensibly with some help from DuBois. The final words of that resolution were "God speed in his noble efforts."

DuBois had been named director of the council's business bureau. During a press interview, he assured those present that the failed resolution, the venomous assault on Washington, did not reflect the sentiments of the delegates in general. In DuBois's opinion, Washington was one of the greatest men of the Negro race.

I had never read of this incident before seeing it in Thornbrough's wonderful book on T. Thomas Fortune. But, I think it is little dealt with in most writings because it would cause a softer landing by Washington-phobes determined to smash him. DuBois worshipping at a Washington shrine could not be tolerated; especially, by the same DuBois who four years later would denounce him in the essay derisively entitled "On Mr. Booker T. Washington and Others?"

Question: Why do you think DuBois suddenly turned?

Answer: The question I asked was, did the real DuBois go into hiding after that? But, to give my opinion as to why the change, please remember that DuBois had been appointed director of the National Afro-American Council's business bureau at the Chicago convention. In preparation for the 1900 council convention scheduled for Indianapolis, DuBois held a conference of businessmen and others at Atlanta University to draw up plans for a national black business organization, which would be presented in Indianapolis. But just before the scheduled council convention, Washington called an unscheduled conference of Negro businessmen in Boston.

At that conference, the National Negro Business League was organized. Booker T. Washington was elected president and held the office until his death in 1915. But, to

Questions and Answers: Part 2

Washington's credit, according to David Levering Lewis in his book on DuBois, Washington did acknowledge that the concept for the business league was DuBois's. But, I think that might have been too little too late for a man with an ego the size of DuBois's. His star was taking on its own characteristic sparkle.

DuBois needed no assistance from anyone on his rise up the ladder of distinction. The Philadelphia sociological study had several years earlier placed him in company with America's towering intellectuals of his time. So Washington's embarrassing him by negating the good work he had put into planning a national Negro business organization reverberated through America's Negro community.

Question: Do you have any opinion as to why Washington stole the jump on DuBois?

Answer: I think that Washington more than anyone during that era did not see in DuBois the talent required to lead and direct people. I suspect that Washington was not satisfied with the election of DuBois as business bureau director of the council, fearing that he would be selected to lead the organization. Washington was perhaps comfortable with DuBois as a developer and planner but not as an executive or strategist.

DuBois's subsequent failed leadership with the Niagara activity bore out Washington's estimation of the man. The Niagara activity shows DuBois at his worst: a bungler in the role of executive, his charismatic sheen fading after the initial meeting evidenced by a serious depletion in the ranks of the original cadre of 29, due to the same repeated protests and rantings against Washington's policies.

Samuel R. Spencer, Jr. in his book, *Booker T. Washington and the Negro's Place in American Life,* writes: "As DuBois admitted, some of the criticism of Washington stemmed from mere envy and from the disappointment of displaced demagogues and the spite of narrow minds. Even DuBois's attacks showed a distinctly personal element." When his Niagara activities are analyzed, he falls far short of meeting the standards required to be a leader of men.

His years at the helm of *Crisis* magazine were DuBois's best years and saw him at his best. With supervision and guidance, he could have been more than just a reporter, essayist and chronicler. Under better guidance from DuBois, black America might not have waited until 1933 for a book on the order of *The Mis-Education of the Negro*, which Carter G. Woodson published, that laid bare the intellectual dishonesty of the few college trained blacks and the causes of black academic paralysis and economic stagnation. Then DuBois, not Woodson, would have had the distinction of writing the second of only two books since 1865, the first being *Up from Slavery* with implied solutions for modifying black lifestyles and achieving racial and cultural improvement.

And, like Woodson, DuBois might have spared the readers the agony of names of individuals, who because of their envy and jealousy of Washington, committed their total energies to despoiling him and his image without once offering a better idea replacement for what they unwisely were bent on destroying.

Question: Wasn't DuBois more committed to advancing his Talented Tenth concept?

Answer: DuBois's Talented Tenth pronouncement more than anything else during his many years defined him. He was given to many theories and notions but could give no hint as to how to expand them for more practical minds to make real. As I have said regarding the Niagara affair, many left DuBois's conferences informed, but only few left inspired. He stated the problem magnificently but only generalized about a solution. His Talented Tenth preachment anticipated stating the problem to the aggrieved and saying to them, "Now you just sit tight, and we'll solve it for you."

To problems created by the anti-rights-for-blacks juggernaut white America set in motion after the 1875 Civil Rights Act was declared unconstitutional, he did say in essence, "Sit tight, the Talented Tenth will agitate against them." Washington's advice to solvable problems was simple and direct: "Let's work on the solution together." To unsolvable problems: **"Develop yourselves qualitatively so that you become indispensable to the needs and ambitions of your community."**

There was the challenge for DuBois's Talented Tenth: to develop black America qualitatively. Ah, but to have pursued that course would have required that attention be directed away from DuBois, and he could not have abided in any distraction away from himself. That self-centeredness is characteristic of many of today's black leaders and is the most formidable barrier to making the black effort a "we" struggle. That fault is fodder for the dependency syndrome that prevails now in the ranks of the black poor.

Question: Is the Talented Tenth idea worthy of pursuit now?
Answer: Let's compare. Washington declared economic power as necessary to the unequivocal acceptance of blacks as a desirable weave in the social and cultural fabric that is America. As the basis for making that scenario a reality rather than empty words, he organized the National Negro Business League to facilitate black business growth.

When DuBois's Talented Tenth essay was published, there were about 12 million blacks in America, but there were only about 2,600 with college degrees. The challenge to DuBois was to announce a plan to increase the 2,600 to 1.2 million by, say, the year 1920. But, that announcement was never made.

DuBois not only heaped many invectives and vituperative on Washington's person and image and otherwise tried to discredit him for many years, but in his essay entitled "The Talented Tenth," to a highly vulnerable black readership, he boldly attacked that part of the Atlanta Cotton State Exposition speech that said, "Ignorant and inexperienced, it is not strange that in the first years of our new life we began at the top instead of at the bottom; that a seat in Congress or the state legislature was more sought than real estate or industrial skill...."

DuBois contended that never in history has culture filtered from the bottom up but from the top down. If his few followers had had their thinking cap on then, they'd have reminded him that at the bottom Negroes would be in the right place to catch some of what filtered down. Or, they would have cautioned him to back off that position because Christianity, Muhammadanism and Buddhism did not filter down from courts of rulers to the anxious masses.

Rather, it emanated from grassroots through the masses across the moats, up the castles' walls and into the rulers' courts and overwhelmed all in attendance. And, if he had lived a year or so longer, DuBois would have witnessed a political and social transformation in America that originated in the steamy Establishment Induced Black Oppression Center (EIBOC) of Montgomery and filtered up to ultimately overwhelm the world.

There are about 2.2 million degree-bearing American blacks according to the 1992 U.S. Census Bureau, roughly 6.2 percent of America's 35 million black population. The Talented Tenth issue should be placed on the front burner

Questions and Answers: Part 2

if it becomes an interim goal, that is, to increase the number of blacks with degrees to a tenth of the black population by a certain time. While attainment of that level would still put us at roughly half the national standard, we would be making progress. But, even to reach the ten percent level, the American Black Community would have to put considerably more collective effort than it ever has committed to academic or any other black efforts.

Question: You earlier noted that black history is being profaned into revisionist history. Are you not contributing to the problem you profess to be correcting? Aren't you also revising black history?

Answer: I have not denigrated DuBois's brilliance as an educator, nor as a writer or chronicler. He was that rare genius with irrepressible energy. He had a penchant for being where the action was, and debating was the order of the day. He was always in search of an audience for his erudite rhetoric. Rhetoric, understand, not oratory or statesmanship. He was well-equipped with words, words, words. But, still no one quotes him. Yet in "Revolution," an essay in his book *Dusk of Dawn*, he boldly claimed that in the period 1916 to 1930, he was the catalyst that revolutionized the attitude of the American Negro toward caste, which made them aware of themselves and what they were capable of and edged them on to self-assertion.

DuBois contended that common slogans Negroes were using in 1940 had their origin in him. Far be it from me to deny a superannuated senior from indulging occasional flights of fancies but having experienced Charleston, Washington, New Orleans and New York societies in 1940 and knowing the tendencies of my people then to emphasize or play up the extent of their non-blackness, I award his memory on the caste issue a patronizing smile with no comment. As to the common slogans claims, they might not have reached the Golden Gate, Renaissance, and Savoy Ballroom before I left for other hip haunts for jumpin' jivers. Here, he is conning historians.

Even DuBois's claiming that he left the NAACP in 1934 because of a difference with management over its racial policy and that it opposed his advocating that Negroes should improve themselves through making their segregated institutions—churchs, schools, businesses, etc.—the epitome of efficiency. This was warmed-over Washingtonism. If that indeed were the case, I should say there was sheer hypocrisy in him and his followers for not going after the NAACP with the same viciousness they went after Booker T. Washington for being at variance with what they thought was a more useful strategy for black progress.

I am aware of DuBois's towering intellectual prowess, and he deserves a place in black history but not as a civil rights activist. He was without focus. He was without an agenda.

His place in history should be determined from a range of perceptions. I don't agree with a physician-friend who said, "I can't for the life of me understand what we are trying to do with DuBois; he didn't do a damned thing other than write and talk as the spirit hit him. Fisk University rightfully has a statue of him as its most notable alumnus, but there are no schools named for him, no streets, no other institutions. That should tell you something."

As a scholar and educator, he belongs alongside the John Hopes, Robert R. Motons, Benjamin Mayses, etc., and as an editor and writer with the P.B. Youngs, Robert Vanns, and the like. As a champion of Negro rights, there is neither organization, nor speech, nor plan of action, nor accomplishment to commend him to posterity.

Questions and Answers: Part 2

The most memorable line from the *Crisis,* which he edited for 20-odd years, was not written by him but by Mahatma Ghandi in a letter to American Negroes in the July, 1929 issue: "There is no dishonor in being slaves. There is dishonor in being slave-owners.... Truth ever is, untruth never was." And, the most candid assessment of Negroes before Washington's Atlanta 1895 speech and Carter G. Woodson's epic book *The Mis-Education of the Negro* was in a November 14, 1891 editorial in *The Age* newspaper by editor T. Thomas Fortune: "The Negro's greatest fault is being a magnificent sufferer."

Emma Lou Thornbrough also credits Fortune for having written in 1883, "We do not counsel violence . . . we counsel manly retaliation." And, Fortune continued in the same mode, "We do not counsel a breach of the law, but in the absence of law, in the absence of proper police regulation, we maintain that the individual has every right, in law and equity, to use every means in his power to protect himself."

I suspect that some of these early 1880's statements might seem to the reader to have been lifted from the 1960's about 80 years later. Well, later that same year Thornbrough again gave credit to Fortune for saying, "When colored men are assailed, they have a perfect right to 'stand their ground' . . . if they run away like cowards they will be regarded as inferior and worthy to be shot; but if they 'stand their ground' manfully, and do their share of shooting, they will be respected, and by so doing they will lessen the propensity of white roughs to incite to riot."

So it can be deduced that Malcolm X was neither a first nor was he unique and that he gained high nationality visibility on blacks' ignorance of their history. At best, he was what he started out as: a religious missionary for a religion

that at the time was little known and little heard of by nearly 100 percent of American blacks. My point is we should ask ourselves the question: What cultural damage has our ignorance of our American black history wrought?
Question: Haven't some black leaders, except Washington and a scant few others, done anything right in your estimation?
Answer: We have used a plethora of recommended strategies except Washington's. If the other strategies had worked, the American Black Community would not be the ocean of malcontents that it is now. American blacks have not learned to differentiate between "rights" and "power."

In 1943, legendary right wing columnist Westbrook Pegler wrote that if he were a Negro, he would live in perpetual agitation. He was grossly out of step with black America's best known conservative, Booker T. Washington, who felt that better use of black time could be made than agitating. The goals of agitation were very limited. Agitation would not prop up the black community that was succumbing to the ills of inattention, intellectual and cultural deterioration or rot, consequences of having failed to water the roots. I will not answer your question directly until I am sure that we have the courage and fortitude to face unflinchingly a reading from the list of our own failings.

How many of us really understand that Martin Luther King demonstrated for rights, not power? You agitate or demonstrate for rights; you fight for power. Failure to make that distinction led to black America's misunderstanding of what King had accomplished and threw the American Black Community into a state of disorientation. "Black Power" and other comedic buffoonery conveyed the notion that the Civil Rights Act of 1964 also

Questions and Answers: Part 2

transferred power. That was evident in blacks' contempt for the black capitalism program.

Few blacks understood black capitalism as grounds on which the struggle for power could be waged. In that regard then, only the National Business League was structured to provide leadership in such a struggle—a fact that did not sit well with reactionary rights groups.

In early years of agitation, Thornbrough notes that Fortune conceived the idea for a national rights organization in 1884: "Let's agitate! Agitate! Agitate! until the protest shall awaken the nation from its indifference." In 1887, Thornbrough says, Booker T. Washington wrote a personal letter to T. Thomas Fortune expressing his support of the idea: "Push the battle to the gate, let there be no holdup until a league shall be in every village." **I think Washington did not favor the Niagara club because agitation without a follow-up, goal-oriented aggressive plans would reap negative results.**

Since the NAACP was going to be started up by whites, the administration would probably be white, and in all likelihood it would ignore the basic needs of Washington's people: **a systematically and deliberately developed Negro business base**. Equally as important was the need for not just a Negro leadership that laid bare to America in general the injustices to which Negroes were being subjected **but also a Negro leadership, which laid bare to American Negroes in particular their counter-cultural and counterproductive indulgences that had to be modified to mirror them as a socially adaptive people.**

As history had already shown, civil rights could be legislated but power could not. King caused "rights" to be legislated as equally accessible to everyone, which included the right to succeed and the right to seize power, a very significant bonus. I think Washington believed too that the

NAACP would reduce some of the influence he and other blacks had among their people, that even if blacks gained control of the organization, if financial support came primarily from whites, the decision on how far they could agree on certain sensitive issues would be dictated by its majority benefactors. There was an even more significant issue that perhaps disturbed Washington: the wrong message was being sent to blacks by some black artists.

In 1900 James Weldon Johnson and his brother, Rosemond, had written the song "Lift Every Voice and Sing" that was being adopted by Negroes as their anthem. Washington, a practical thinker, perhaps saw in the song the *one dangling phase* that even now should cause it to be either dropped as a black cultural element or amended. It is "Let us march on 'til victory is won," a phrase which, embarrassingly, is still being subscribed to by blacks 94 years after it was unfortunately first written: *We **march** to battle* and ***fight*** *for victory.* So soon were we being conditioned for a "hand-full-of-gimmes-and-mouth-full-of-much-obliged" reputation.

Rights agitators react and protest; power seekers aggress and build. Black America needs both in varying proportions, **but it has had comparatively too few seekers of power.** Some equate numerical voting strength with power, but comes election they find that power seriously diluted: of the many registered, few vote. And why? The winners can only legislate more rights but not one particle of power. This validates Frederick Douglass's great lesson that power concedes nothing, never has, never will. It must be seized.

Marcus Garvey was on the right track with his industrial program, but he too was in over his head—a wet nurse to a condition requiring an experienced surgeon. A. Philip Randolph's gamble paid off in his getting what he went after, which increased the chances of success for a power grabbing strategy. Alonzo Willis, a black publisher, cited the Negro community's failure to create value-added businesses during the great influx of money into the American Black Community from 1941-1946.

He told me the food, sweets and clothing industries were where we spent most of our money, and we should invest our monies in building those kinds of businesses in Negro communities all over America. That indicates to me now he was well schooled in Washingtonism. Martin Luther King's agitations reaped great bonuses far in excess of desegregated public transportation.

If Thurgood Marshall's Herculean victorious undertakings and A. Philip Randolph's excessively profitable high stakes pot he raked in with the Franklin D. Roosevelt gamble can be set aside for special study at a later date for another book, then it is easier to say without contradiction that Marcus Garvey's and Martin Luther King's crusades were the only truly successful major thrusts by a large number of blacks since the American Black community came in existence in 1865. **And the reasons for King's and Garvey's successes are plain and simple, yet are not widely discussed: they were mandated by the people they purported to represent.**

From 1919 to 1921, when the average yearly salary for the American worker was $1,180—less for blacks who have always lagged behind average pay—nearly 12 million blacks invested $10 million—that's $70,405,000 in 1994 dollars—in Garvey's Universal Negro Improvement Association. The attraction of Garvey's crusade was aggressive action, the sweet sound of building, an activity mostly alien to his many critics. There is an added dimension to the Garvey story: **He was acting on the legacy of his hero, Booker T. Washington.**

After Garvey's empire fell in the early 1920's, as though to avenge his defeat, Negro entrepreneurs commenced to take advantage of the economic boom of the decade. Vishnu Oak, in his 1949 book, *The Negro's Adventure in General Businesses,* recounts this golden age of black economic assertion when value-added businesses included manufacturing, sports franchises, hotels, movie houses, movies, lumber processing, etc. If Garvey were the inspiration for that economic activity, and Washington was Garvey's inspiration, then to whom goes the credit for the black economic awakening of the 1920's? Unfortunately, the world's economic bubble burst in 1929 and brought an end to black economic awakening, a mode it is still struggling to regain.

King's agitation resulted in a civil rights jackpot that figuratively broke the bank. His condition for accepting absolute leadership of Montgomery's black community must have been, "I talk. We walk." Had the community not accepted, he could not have led it through the final stages of that "severe American crucible" Washington referred to in his June, 1896 Harvard speech.

King's crusade bore bountiful fruit that culminated with America's law books shedding itself of all racially discriminatory and racially restrictive laws, including the enfranchisement of all Americans. **But, there is irony to the King story: there was less than satisfactory black financial support for King's activities.** But, he had the full support and authority vested by Montgomery's 30-odd thousand people.

King had on a local level the kind of black support that Washington implied in Atlanta in 1895 was needed on a national scale if the black community were to make measurable progress: support given no black leader since King.

Questions and Answers: Part 2

No black leader since has evidenced that he can bring anything to the black struggle beyond rhetoric. No black leader has been able to describe goals in specific terms and offer a set of plans for attainment of those goals.

Garvey did not generalize. He defined his goals and convinced over a million blacks that these could be attained only with their generous support. King defined the goals toward which Montgomery's blacks would collectively strive and convinced them that the interim inconveniences they might have to suffer were a small price to pay for the magnanimous rewards inherent in the goals they sought.

Question: The word is that the National Business League's support is shrinking. If total organization of the American Black Community is contingent upon a functional league, doesn't a crisis exist?

Answer: No. As long as there is a substantial black business community—over 450,000 at the last U.S. Commerce Department report—the league's existence should be assured. It should be every black American's responsibility to encourage black business membership in the league. Each black should be taught the role and significance of the league to the empowerment of black America. It is reasonable to assume that a majority of black businesses would contribute an average of $50 yearly to keep the league functional and viable.

It is also reasonable to assume that most of America's nearly 11 million black employees would contribute a minimum of $60 yearly each to a foundation established to move black America into the arena of power as a player and from the fringes as supplicant and alms seeker. That would command universal respect for American blacks as a people of purpose rather than ones in an arrested development as chronic complainers.

The legacy of the DuBoises is to feed from the trough of white foundations and resources of the majority business establishment and government for our most trivial needs, giving no thought to the psychological damage we also were inflicting on succeeding black generations and steering them away from self-reliance and damaging the self-esteem of too many. Many of the problems of black youth can be traced to our peddling of ersatz as the real thing, teaching a history constructed as we wish it had been. Washington taught us the formulae for growth. Garvey tested them, resulting in millions of blacks endorsing his generalship that scared the daylights out of the Establishment.

Black entrepreneurs from 1921 to the ill-fated 1929 expanded on Garvey's methods and created a golden age in black economic presence. Randolph employed the numbers concept that gave President Franklin D. Roosevelt nightmares, and Martin Luther King used the authority invested in him by numbers and reshaped all of America. Every one of these heroes evidenced awareness of Washington's great lesson: **never permit grievances to overshadow opportunities.**

Try to imagine that black leaders are leaders of 35 million blacks, not just of an organization, with a mandate to speak for black America. What has been accomplished in such an arrangement? Total communication in the black community is established. Maybe a black youth reclamation is in progress and at long last racial goals can be set. The "Black Enterprise 100" in a while will be expanded to the "Black Enterprise 1000."

Question: Give some examples of racial goals.
Answer: A national black monument in Washington D.C.;

a national black administrative center in Washington; a great black cultural center in the geographic center of black America that would be no more than an overnight drive for 80 percent of blacks, where black families can vacation and indulge their culture and heritage and whatever competitive sports are being held; three million blacks with degrees by a certain year, up from the present 2.2 million; and, ninety percent of middle school graduates continuing to high school, with an 80 percent graduation rate and 80 percent of those continuing on to college.

There would be black college graduates thinking black businesses as credible sources of employment; a multi-billion dollar foundation; and, every DuBois type in black America ensconced in a niche where he fits, after having simply let down his buckets where they were; religious institutions pooling their resources to form stock corporations; fraternities and sororities merging in Pan-Hellenic corporate ventures; black industrial and high-tech firms girding the world; and, one black business for every 20 black population count versus one for every 80-odd now.

Question: How would black youth benefit from all of this? How would reversing the course too many are now travelling be accomplished, if it will?

Answer: By prescribing and teaching **full personhood** as society's highest and most desirable goal. Everyone must be informed of that concept so that it becomes endemic to the total life of every black child and adult. It is called the **full personhood principle.** It could be incorporated into our daily habits as a ritual of sorts. A parent would ask her/his school-bound child: "What is your ambition?" The child would reply: "To attain **full personhood**." Then, the par-

ent would ask: "How do you attain **full personhood**?" By following the **full personhood principle** that says
Studying is learning,
Learning is trying to know,
Knowing is knowledge,
Knowledge is what makes one whole;
Knowledge is personhood fulfilled.

Question: Isn't that an encroachment on religious principles?
Answer: "That's a "yeah but." A song I know has these words:
Yeah but, yeah but,
Always there, yessiree
That simple yeah but,
That yeah but psychoses
It's always yeah but
Said oft to disagree,
To disrupt harmony,
For continuity
Or be a disgusted.

Anyone construing the **full personhood principle** in a religious context surely has problems with the Pledge of Allegiance and the national anthem and is in dire need of special attention.
Question: Aren't there lead-in instructions for the **full personhood principle**? You can't just start in exacting compliance or requesting adherence cold turkey.
Answer: If I were introducing the principle to an adult or child, I think I'd first explain that in order to fully understand it, **one must be able to differentiate between seeking to get attention and seeking recognition.** Attention-

getting, being a form of buffoonery, has absolutely no place in the lives of subscribers to the **full personhood** principle. Recognition is a kind of reward. Please remember, the **full personhood principle** is intended for the positive development and growth of a person. When I read through the myriads of essays of DuBois because of his top-of-the-line education and training in education and social studies, I expect to stumble upon such guiding principles at which occurrence I can say triumphantly "Eureka! I've found it; I've found DuBois's legacy!"

Question: If your assessment of DuBois is correct, if he has in fact been oversubscribed, what damage has been done? Are we any the worse?

Answer: The damage is incalculable, the extent never to be known. Worse yet, the intellectual honesty required to attempt damage assessment does not exist in black America. But, of the three classes of opinions—expert, learned and respected—mine would have to be a respected one. I'd be given the courtesy of a hearing in deference to my seniority. Leaders take stock of their resources and draw up plans for impending engagements.

After DuBois broke with Washington, DuBois was thrust into a place in Negro America's catbird seat, which he had neither earned nor deserved. From my perspective he immediately realized disorientation, dysfunction, and an inability to produce. Miscast as he was, with no notion of how to use the resources available in the black community, DuBois was therefore unqualified for drawing up a plan for Negro assertiveness. Unable to build or nudge black America forward, he resorted to the time-dishonored tactic of tearing down. He committed his writing abilities to besmirching Washington's image, a task at which he became America's foremost expert in time.

DuBois more than anyone else is responsible for black America's propensity for self-deception. He did that by overselling himself to a gullible black intelligentsia. When he advanced his Talented Tenth nonsense (isn't it standard that the exceptional persons from any group, who rarely equal a tenth of the group, define the group and give it direction and purpose), he figuratively crowned the head of every black college graduate for years to come. I've heard speakers at college commencement exercises and fraternity conventions stroke and massage egos, as DuBois did, by addressing the graduates and conventioneers as the Talented Tenth—the untested and unproved, the worthy, unworthy and worthless.

Read Booker T. Washington's Atlanta speech and note that he seemed consummately and fully cognizant of the resources available in the Negro community for the conflict to be waged and that he spelled out how best to appropriate those resources in a carefully thought out strategy for Negro development and growth. Even now a functional black America can be constructed from the plans Washington unveiled at Atlanta. DuBois left no plans, not even a theory of education, which he touted so highly and accused Washington of holding in contempt. What damage has been done? Far too much happened from 1900 when Washington organized the National Negro Business League to the founding of the National Association for the Advancement of Colored People in 1909 to pass the test of innocence.

Question: This is a new twist. Has there been some dust swept under the rug?

Answer: It's only a theory. But, DuBois might have been an unwitting tool used by the white liberals who organized the NAACP to undercut Washington's considerable influence. Most of them had attended Afro-American Council conventions and observed how strong was Washington's influence even in his absence. They had no cause for concern when he was just teaching the gospel of economic power but to have organized the National Business League to orchestrate the building of economic power would eventually shut the whites out of the process of managing Negro affairs.

Monroe Trotter, editor of *The Guardian*, viciously anti-Washington, perhaps influenced DuBois to express views that contravened Washington's. The strategy was to mollycoddle DuBois and keep his ego massaged (an incorporator of the NAACP, a staff position, etc.). The irony of it all is that the reason that DuBois was brought into the NAACP—to snafu Washington's economic strategy—was the same one that DuBois said caused him to leave the organization in 1934: **teaching the decades-old Washington concept of developing Negro economic power.**

Question: It causes one to wonder, but then when the anti-Washington essay is remembered, your theory becomes a bit farfetched.

Answer: It is the alleged anti-Washington essay "On Mr. Booker T. Washington and Others" in which this theory has its roots. Criticism of Washington was guarded, not vicious and scathing. It was restrained, almost apologetic. I even detect a bit of timidity but none of the arrogance so characteristic of DuBois. Many white liberals of that period used extreme caution in references to Washington, not wanting to offend Negro America's liberal elements.

A big plus for the white NAACP group was Mary Ovington, whom David Levering Lewis in his book on DuBois labels as one of "the sparks of the association." Mrs. Ovington and DuBois had a mutual admiration for each other. She was ever timely in praising his works, especially his "Credo." In addition, Dr. Lewis points out, Ovington raised money for Atlanta University. There were many other subtleties employed to keep DuBois in the fold as the only voice in the group that could speak out against Washington and not convey the notion that the whites and the NAACP, with invaluable assist from the indispensable Monroe Trotter, were really directing the show.

We are now a people with indefinite allegiances, who are highly vulnerable to bearers of better ideas, a concept advanced by DuBois in his break with Washington to strike out on his own. If once we strove to hone and fine tune our American culture that has been put on hold, now we struggle with totally alien Afrocentrism under assault from some claiming to enlighten us as to who we are.

Since time began, religion has been the centerpiece and defining element of any culture. Any attempt to superimpose another religion on a culture fractures that culture beyond repair. For better or for worse, Christianity is the centerpiece of the culture of black Americans. Afrocentrism and Islam represent the "go back" in the ages old counsel of the sages: "It's all right to reach back, look back, and even give back, but never go back."

These distractions from our normal and logical pursuits are possible because blacks have been deluded into believing black America is engaged in a struggle. Recently added to our opposition in the struggle is the Jew. Since when did the Jews and blacks become racial adversaries? We know they have no history of religious conflict—black Christians versus Zionists. We accepted the legal services of Joel Spingarn with the NAACP from its inception and now honor his memory with an annual medal bearing his name.

Questions and Answers: Part 2

Joel Spingarn notwithstanding, we know that Jewish lawyers defended the nine Scottsboro Boys in 1934. Jack Greenberg was one of the lawyers who assisted Charles Houston and Thurgood Marshall to plan strategy for *Brown vs. Board of Education*, and Martin Luther King, Jr.'s right hand and one of his principal advisors was Jewish. Just recently I read in a local paper where a young black adult complained that Jews designated who would be black leaders. He should demand a voice about such designation proportionate to his financial support of black leadership.

Question: Who pays the tow runs the show, right?

Answer: Booker T. Washington's consummate fear was that blacks would learn that lesson too late. Time has conferred justification on that concern. **We have pitifully little of what gives a race a presence of substance in a society. Martin Luther King, Jr. caused the playing field to be leveled, but black America has never fielded a team. Black America's most crying need is a commanding presence.**

Yet, it has passed up many opportunities to build such a presence, simply because a benefactor **outside** the black community could not be found to fund whatever it took. So that need, a crisis of the first magnitude, still goes begging. **Our personal resources seem to be off limits where providing for our own social, cultural and economic uplift is the goal.** History is replete with examples of people who during extreme crisis rise above themselves to effectively and successfully handle the crises but such examples don't scream out to us from the pages of black history.

Question: Aren't you presuming that all blacks are conversant with black history? For instance, how many blacks know about King leveling the playing field but black America having no team on the field? How many blacks know the benefits that accrued to us as the result of A. Philip Randolph getting Roosevelt to issue Executive Order 8802 in June, 1941? It seems we might have an ignorance-of-black-history crisis, wouldn't you say?

Answer: That's part of the problem, but it's worse. We have a leadership crisis. Since Washington's death, black America has lived through a period of non-innovative, non-creative and less than imaginative leadership. **DuBois, who hailed the Marcus Garvey projects as a "mighty coming thing," failed to recommend "lessons learned" sessions by black intellectuals to determine what in Garvey's program caused blacks to invest the unheard of astronomical amount of $10 million for use in future business ventures to improve their success ratios. Instead, he used good *Crisis* magazine space later to detail the negatives of Garvey's program, which caused its failure.**

Nor were "lessons learned" conferences held on the heels of Randolph's and King's successes. Black America has grown comfortable in the undignifying posture of alms begging, a more common expression of wanting everything for nothing as evidenced by the incessant pursuit of foundation and Establishment money that defines most of the high visible black leadership. The monetary demands made of the black masses go largely unanswered because the purposes given are generalized, indefinite and will bring credit to a personality and/or organization if the dice throw comes up favorable.

Question: Isn't that traditionally the way it is, that some get the credit?

Answer: Then let's be consistent and pass out some discredits. Let's have some heads bowed in shame and embarrassment. Who should be discredited for our having only one-third of our proportionate share of college graduates? Conversely, who should be discredited for never having apprised black America of the great strides it has made in education? Improving the education formula becomes a racial problem, requiring a racial goal. **Individuals don't attain black racial goals. Black Americans do.**

Question: Your implications are that things are not going to get better until we are all pulling together in the same direction toward a common goal, right?

Answer: Not toward **a** common goal but toward common **goals**. Presumably, goals have been set, if any have indeed been set, with funding by Establishment sources in mind. The ideal situation will be for specific goals to be proposed as being needed for black America, and public discussion of the proposals conducted. The next goals should be material ones: the 130-odd-year overdue black monument in Washington, D.C., the black America administrative center and the black American foundation building.

Two of the most uncomplimentary comments about black America came from an Asiatic Indian at the Memphis Airport in 1969 and an American white young man about a half dozen years ago. The Asian commented that there must be upward of five million blacks in America. When I told him there were in excess of 25 million, he registered extreme surprise before examining the Memphis skyline. Then, with a wave of his hand to indicate his area of focus, he commented: "Your word, sir, but it doesn't show; there is no evidence of there being that many."

A young white boy was thumbing through a *Black Enterprise* magazine annual issue of the 100 largest black businesses in America In the waiting room of a dental office. He leaned toward me and said, "Sir, aren't these really the 100 largest black businesses in Georgia?" When I assured him that was the list of the largest black businesses in all of America, he looked me squarely with anger-filled eyes and said, "This is disgusting, enough to make one sick. I'm floored!"

On another occasion, a young black police officer turned from a story on the coffee shop television to say to me, "When is anyone going to observe that these black boys are not really bad, they are simply bored and relieving them of that boredom is black people's problem?"

Question: Did you take all of these observations into consideration before you concluded that black problems are solvable?

Answer: Those and more. For instance, you can't be a freewheeling Christian at lunch, at dinner renounce Christianity for Islam and be an incurable, genuine Jew baiter by breakfast. That defies logic, is immature and anticultural. It also gives black presence a nuisance value. And, it is typical of the confusion introduced into black America since the 1950's by individuals with "better ideas" of what is culturally endemic or correct for black Americans. The traditional gullibility of American blacks does not help in countering the assaults. Cultures are founded on conventions that became endemic to our lives and take ages to develop: religion, art, names, music, national identification, national dress, speech, etc., which is collectively called cultural heritage.

The cultural heritage of our forefathers underwent merciless assault and transformations dictated by whites after they were displaced into slavery. Now after centuries of modifying their culture into a very distinctive and unique American culture, blacks are under another merciless assault from within, to modify again, with no notion of where

Questions and Answers: Part 2

it will all lead or what will result. Blacks are in dire need of just such distractions as I have been able to construct based on my exposure to Booker T. Washington's great teachings.

I never had any doubt about who I am: I am an acculturated black, a uniqueness I share with tens of millions of others. I am satisfied with that. But, even if I knew of my specific ethnicity—of Nigerian, Rwandan, Ugandan, etc. descent—how could I use that to improve my present status? What advantage would accrue to me? Rather than modify my acculturated status, I opt to use my fading energies to steer the myriad of modern DuBoises caught up in the frenzy of misinformation into more constructive, productive, useful and purposeful pursuits.

These pursuits include casting down their buckets where they are and resetting their compass on a new reading, a new course from the present one, which is unarguably leading to racial madness. Or, have us follow more productive pursuits as learning what it takes to be an outstanding American citizen and apply what is learned to becoming just that.

—*Five*—

A Capital Accumulation Plan for Black Businesses

I am an advocate of black business organizations, having founded one in 1968 in Hartford, Connecticut and managed it until 1972. I have been witness to the great presence to which the Hartford black community was transformed because of a business organization's traditional function.

I have had the good fortune of hearing passionate narrations on Booker T. Washington by Berkely G. Burrell, then president of the National Business League, which was founded by Washington in 1900. I think time will bear out that Dr. Burrell, who died in 1981, was the spark in the 1960's and the 1970's that fired black youth's interest in business administration and caused many black colleges to expand their business courses.

My feeling is that every black business should be a dues-paying member of at least one black business organization—either local, regional or state—with administrative ties to the National Business League. If there is no local organization, then a regional or state one should be available. More importantly, there should be no **competing** business organization. **The National Business League is part of our heritage.**

It is the oldest black organization in existence. It stands now as a test for our commitment to our heritage, and we must keep it viable, vibrant and functional. To allow it to fade into oblivion would be equivalent to presiding over the demise of Tuskegee University or Bethune-Cookman College.

The launching pad from which the American Black Community moves forward politically and economically is its black business establishment—all 450,000 black businesses organized as one, not fragmented into scores of different organizations.

The black business establishment is our only power base and properly appropriated or directed it can leverage many benefits to our side of the ledger. Black America cannot yet afford the luxury of a fragmented business establishment. I don't mean, however, there should be no homogenous business associations like banks, drugstores, mortuaries, builders, etc. Business purity is also highly desirable. Black businesses have a social responsibility that often can best be discharged in concert with other business. To that end, the black business organization is the facilitator, assistant and coordinating agent.

The black business organization is a lobby and represents its members in influencing legislation favorable to its membership and to the community the membership serves. If the membership supports the organization so that it remains membership-supported, then the organization can progress as much and as far as it has to for the membership and the community. Otherwise, in instances where it must seek outside funding to remain functional, it can only progress as far as that outside support allows it to. You've heard it before: Who pays the tow runs the show.

I see very slim chances for any appreciable cultural, political and economic growth of the American Black Community until a significant majority of black businesses is part of an organization scheme as I have described.

Smoke and fumes from black-owned businesses do not make measurable contributions to the smog conditions in American towns and cities. That hypothesis is not an environmental plus for which we should be proud to accept praise. It is a disastrous economic circumstance from which we must hasten to extricate ourselves.

The years since 1967 have been for me a period of almost total dedication to black business development, which began with my being part of a cadre that founded and built a black-oriented bank. **My initial position during that period became steadfastly this: If black America is ever going to have power commensurate with its size, its black business establishment must also be proportionate with its size, And, if that business establishment is going to become a reality, some time the start-up capital for building and expanding businesses must be provided by the American Black Community.**

The business base is the foundation of power, and white financial institutions are not going to fund us into power. Alexa Henderson in her book that traces the history of the Atlanta Life Insurance Company noted that the main targets of white rioters in the 1906 Atlanta racial disorder were black businesses, which they called "the eyesores of black progress." But, having to underwrite the cost of some of our own economic growth should not frighten us. We have the capital as I shall show later. But, do we have the

A Capital Accumulation Plan for Black Businesses

commitment, the intestinal fortitude? I think we have all that is necessary. Otherwise, I would not offer this plan.

I think one of our problems is having among us far too many unmandated talking heads who feel that the black masses are incapable of making their own statement and that others must agitate (make statements) for us. They learned nothing from Marcus Garvey's activities, A. Philip Randolph's monumental triumph and the statement made by the Montgomery black community in 1955 that ultimately resulted in the social recasting of America.

Who of us have not heard it said by another black: "We are not ready"? For what aren't we ready? That is an accusatory statement that should be directed at white people who were not ready for a progressive black community like Wilmington in 1898, so they summarily destroyed it and all its black businesses and drove the blacks out of town. Black businesses suffered the same fate in Atlanta in 1906, in East St. Louis in 1918, in Tulsa in 1921 and in Mississippi everywhere all the time, often under the observance or leadership of the police. For what specifically weren't whites ready? For blacks as employers of blacks; for blacks who wore a suit and tie instead of coveralls; and, for blacks who did not show up in the labor pool each morning?

Blacks were ready when Marcus Garvey beckoned in 1919, but whites were not, so they summarily ruined him. Blacks evidenced readiness was the reason Randolph triumphed in June, 1941, and whites later committed wartime nuisances against blacks in Detroit, New York City, Waco, Mobile, Beaumont and other areas of high defense industry concentration. But, blacks served whites notice at Montgomery in 1955 that their unreadiness no longer mattered. Montgomery, blacks have governed, mayored, legislated, chaired, generaled and made Atlanta one of the world's great cities and gave the region in which it is located a measure of dignity and guarded respectability disproportionate with its history. Ralph Bunche, Robert C. Weaver and Thurgood Marshall were no illusions or anomalies after all.

But, what we are to demonstrate readiness for now is not conventional. The task is capital formation for investing in existing black businesses, building new businesses and taking advantage of favorable investment opportunities that might become available.

Booker T. Washington's great lesson at Atlanta in his epic speech there in 1895 was pointed and direct: **"Nor should we permit our grievances to overshadow our opportunities."** That suggests to me our not being organized now is the best reason to start immediately to do so and not waste time regretting that we are not. For until we are organized, we will never be able to maximize our economic and political potential.

The American Black Community consists of 35 million people, about 12-15 percent of the country's total population. Over 20 million are of voting age, an awesome figure in relation to America's 187 million registered voters. **However, ours represents only potential power.** It can't be realized to its fullest measure except within the framework of a **black economic strategy**, which black America has never had or even attempted. So a strategy cannot be developed before the American community is a unified force, **which by its very nature would be a racial statement of fearsome proportions.**

Necessarily, I will describe what could be a massive capital accumulation project using only about five percent of black America's employment force. I don't normally

A Capital Accumulation Plan for Black Businesses

depart from my practice of considering black America as one "body" except to identify the black business establishment. However, before I identify the particular group needed for this capital formation plan, I wish to relate the source for the project.

It is an often overlooked and little-used example of racial cooperation, the profoundest example of concerted action for a common purpose ever written. The story goes that Biblical Moses had gone onto Mount Sinai to confer with God and left his brother, Aaron, in charge of the Israelites who had fled from Egypt. The Israelites became restless and maybe jealous too and told Aaron they also wanted a god to worship. Aaron assured them that they could have a god, but they would have to pay a very dear price: the women would have to throw their golden earrings into a common pile.

In effect, he was telling them that anything of value could not be had just for the asking but that did not mean it wasn't attainable. Sometimes a great sacrifice has to be made. The price for having an idol in this instance was the golden earrings of the Israelite women. They complied. Aaron melted the gold and shaped it into the image of a calf, which the Israelites worshipped. The real story is not the fashioning of an idol but that full cooperation was needed and was given.

One earring could not have been transformed into an idol, but hundreds of thousands of little gold pieces could and did. The surrendering of the earrings was the Israelites' racial statement.

Now is the time for the American Black Community to build its **golden calf.** What do we want?: economic parity, political parity and a sense of hope reflected in the faces of all of our youth.

What is being done now in the way of business loans to blacks leaves much to be desired, and the situation shows no signs of improving. But that has always been the conduct of white commercial lending institutions vis-a-vis blacks wanting to open businesses or wanting to expand existing businesses almost without exception. As I said earlier, whites are not going to finance us into competition with them.

President Nixon's Black Capitalism program was not designed to be a wrecking ball to the barrier between white banks and black businesses. Only a few cracks were made in the barrier. Black capitalism was an eye opener. It allowed us to reflect on our own sordid economic circumstance and showed us we had gone counter to Washington's guiding principle by having permitted our grievances to in fact overshadow our opportunities.

There is a certain mystique about a businessperson that causes the community in which the business is located to accord him or her a fair measure of respect—respect that multiplies many times if the businessperson is an employer. That is a polite way of saying the business is esteemed and revered and wields power. Too many black employers in one community might result in too much power concentrated in black hands. But, that is exactly the posture Booker T. Washington wanted the American Black Community to ultimately assume, since the other half of the power equation—political power—was being made inaccessible to that community.

I hope it is clearer now why if the black business establishment disappeared from the American scene this moment, the American stock market would not react one smidgen. It is that kind of economic irrelevancy Washington taught us not to let develop. At the stage

A Capital Accumulation Plan for Black Businesses

we are now, **we are courting racial expendability.** We have been managed into the economic and political posture that now defines us as a people. We can overcome that condition.

After many years of intensive study of Booker T. Washington's economic and political philosophies, I was able to develop the **golden calf** concept. Only one small **golden calf** will be built this time. By extrapolation, you can contemplate the enormity of the really big one when all black Americans are participants.

There are about 450,000 black-owned businesses in America. Their total annual receipts are $25-30 billion. They employ an estimated 900,000-plus people. No business employs as many as 5,000 people and only a dozen or so employ 1,000 or more. Many small businesses are operated on a part-time basis because the owners are full-time employees in the Establishment. To the mind trained in economics, the figure 900,000-plus conveys a lurid message. But, 18 years ago, the number was 217,000. And, black America was then as it is now in near total ignorance of how effectively their 35 million population community is doing in providing for its own. With 10.8 million American blacks employed, the 900,000-plus employed by black enterprises suggests much more needs to be done.

How much more? Just as much as we can. We will be under international scrutiny: 10 million black American youths and nearly 300 million blacks in Africa's 54 countries and the Caribbean nations and possibly some of the islands of the Pacific.

The total annual income of black America is nearly one-half trillion dollars. That's 5.7-6.0 percent of all income paid by the American business establishment. But, black Americans are 12-15 percent of America's population. I hope you see that black America now is even closer to economic irrelevance and expendability than was suggested above or than we care to admit.

Now let's discuss the capital accumulation project—the making of our own **golden calf.** For this plan, about 500,000 or so easily identifiable individuals—the membership of the eight black Pan-Hellenic (Greek-letter) societies representing four sororities and four fraternities—will be used. Each society operates a national office from which it administers to a network of graduate and college-based undergraduate chapters.

They are the world's largest repository of college-trained and university-trained blacks. Many highly visible and otherwise renowned blacks are members of these associations: educators, poets, novelists, historians, judges, politicians, civil servants, lawyers, businessmen, businesswomen, physicians, etc.

For those of you who are skeptical about using the black Pan-Hellenic societies for such a massive task, ponder this: The Delta Sigma Theta sorority made a **featured movie**, "Countdown at Kusini," and I personally hailed it as Delta's **golden calf!** I further reasoned that if one society soliciting 100,000 or so members could raise enough money to finance filming a movie, then eight societies soliciting their collective membership might be able to buy a movie-making studio!

I also note that the Masons in Detroit had their **golden calf—a television station**—and reasoned again that maybe all black Mason societies might be able to pool their resources and acquire a national television network!

How do these organizations finance their administrative duties each year? As it now stands, the national office of each society assesses each active member of the gradu-

A Capital Accumulation Plan for Black Businesses

ate and undergraduate chapters each year. The taxes are appropriated to cover the expenses occurred by the national office. To initialize the capital accumulation plan, the first step that the eight associations must take is to meet and organize a stock corporation, that for now will be called Pan-Hel, Inc. Initial capitalization of Pan-Hel, Inc. could be accomplished by each association being authorized to purchase common stocks in Pan-Hel, Inc.

For instance, they could be authorized to apply $10 of each member's payment toward the purchase of common stocks. The stocks purchased by each association would represent that association's ownership in Pan-Hel, Inc. Each association would have the option to purchase a certain number of stocks within, for example, a five-year period, say by December 31, 2000, at the same price per share as the initial offering.

Today, at many of the local chapters, each member is assessed a certain amount yearly to conduct that local chapter's business during the year. Mailing notices, conducting social events, stationery, and academic scholarships are some of the expenses local chapters incur. The assessment often is several hundreds of dollars.

A stock option could be offered to active members, say 500 shares at $10 a share until a certain date. The local chapter could also be offered the same option with the same qualifier. At an average investment the first year of $250 per member, plus investment by each local chapter, **black America could conceivably have a business capitalized at several hundreds of millions dollars. Conceivably, in the first five years, Pan-Hel, Inc. could be black America's largest black business ever!**

What would be the purpose of Pan-Hel, Inc.?

It would be the same as any other holding company

with slight variances due to nature of our cause. The accumulated capital is a **golden calf**. The **golden calf** can be molded in one year and proclaimed in triumph, or fattened for four more years before giving it a more functional purpose. Perhaps in the first year or two, investments could be made in short term securities while in the interim business searches are conducted to locate existing black companies with growth potential as well as viable business start-ups.

This approach is not just blue-smoke and I do sincerely hope that many of you reading this idea can understand and see the vision—for without vision, there is no hope. As a person seriously involved in business for a large portion of my life, I fully understand all of the ramifications and immediate objections of this plan. The doomsayers (both blacks and whites) will be the first to poke holes in it and proclaim why black Americans will never be able to produce such a massive demonstration of power and purpose.

Nevertheless, I fully believe our day has arrived. We have the best brains we've ever had, the best institutions and the best business minded people of character. Therefore, I say to each of you—never say never. Rather, say to yourself and other: If not now, when. If not us, who?

Many people have asked me why is the number of blacks with degrees so important to the success of Pan-Hel, Inc? I shared with them a theorem I devised years ago: The size and quality of a culture's business establishment is directly proportionate with the number and quality of the college trained within that culture.

Black America has never had enough college graduates to make a representative academic statement for black America, nor yet to give the American Black Community a

A Capital Accumulation Plan for Black Businesses

presence of substance in academia. My experience as head of the Hartford black business establishment was that as the number of blacks with degrees entered the entrepreneurial ranks, the numbers of blacks employed in the Hartford black business establishment increased. Except in rare instances, all-black *Enterprise* 100 businesses are headed by college trained managers. There will probably never again be another A. G. Gaston, Alonzo F. Herndon or Madam C. J. Walker.

The very nature of businesses today dictates the prospective manager has high academic credentials. More and more lenders are becoming partial to prospective entrepreneurs with graduate degrees. On the strength of my theorem then, this hypothesis logically follows: If with 2.2 million degreed blacks there are 450,000 black-owned businesses, then the business prospects with seven million degreed blacks are at least 1.4 million black businesses. And surely the number of value-added businesses will grow. In time, the *Black Enterprise* 100 listing would have to change to the *Black Enterprise* 500 or maybe *Black Enterprise* 1,000.

While there are no exact or official figures as to the number of employees in the black business establishment, based upon the best data available, I derived that there are about 2.25 employees per each of 450,000 businesses or a total of 990,000 employees. If that same average holds for 1.4 million businesses, there would be over three million employees in the black business establishment when it grows to 1.4 million businesses.

Think of it. Black distribution centers, manufacturing firms and food processing plants could be initially or mainly funded by Pan-Hel, Inc. Instead of our best black minds working for Uncle Sam, they would be working for a new black America!

Businesses attract businesses and businesses beget new businesses. For every new or expanded firm, suppliers are needed. Often, depending upon the size of the businesses, new supporting facilities must be established. Just the presence of Pan-Hel, Inc. for example, would trigger the establishment of paper manufacturers, business forms suppliers and computer firms. The banking and financial institutions would also be impacted.

There would be other positive social benefits accruing to black America and society in general that can't be factored in. For example, black America has no extensive history of substantial economic successes resulting from commercial ventures. Thus, the impact of Pan-Hel, Inc. type ventures on black youth cannot be predicted—but think of the possible impact!

This has necessarily been merely an abbreviated presentation. As for the next steps, I would say it is up to the National Business League. It must decide the efficacy of this project and if it favors the concept, then it must summon the head of each fraternity and sorority and discuss it with them. It would also be advantageous for a business learning center such as Howard University, a business school or a black think tank to conduct a thorough analysis of this approach. But there is a **golden calf** and not a penny need to come from external sources.

The cry of the Israelites was for a god to worship. Their leader set the terms that had to be met in order to provide them with a god to worship. They met those terms and a god, the **golden calf,** was created.

As was done to Aaron, demands are being made of the black leaders by the black masses. They want jobs that pay more than marginal wages. To provide such jobs, we need to build substantial businesses. Blacks want more

A Capital Accumulation Plan for Black Businesses

businesses. In order to have more businesses we need the capacity for large financial investments in those businesses. Therefore, we must do for ourselves what the banks will not do for us. Oh, the banks will come around once they bear witness to the fruits of our resolve. The Pan-Hel, Inc. plan could become the model for similar ventures involving other groups such as brotherhoods, trade associations, professional associations and religious institutions to name a few—each building its own **golden calf.**

In addition, I predict there is a lot of powerful inspiration that will spill onto the blacks of the world as a result of black America inching toward economic viability. In time, the beneficial fallout will be mutual. **I subscribe to the hypothesis that the key to activating the blacks of the world is to activate black America; get black America on the move and the entire black world will commence to move!**

But first, each and every one of us, from the youngest child to the oldest elder, need to understand that the name of game is commerce. Take away trade and commerce and what's left is no school, no church, no library, no parks, no city hall, no people—nothing. There's another way of saying it: **Every town is not all business, but every town is because of business.**

May there be many, many **golden calves** in our future!

—*Six*—

Where Is the Black Press?

As a black business activist, I have always been cognizant of the need for and importance of involving the black press/black media in any major event about to take place within the American Black Community. Learned people know that when it comes to launching a major event and keeping everyone involved informed of what's taking place, a reliable communication apparatus needs to be fully functional.

I believe that an often overlooked but central reason for the lack of any kind of national black economic movement is not because of a lack of ideas; **rather it is because the appropriate channels for the wide-scale dissemination of those ideas to the total American Black Community has and continues to be lacking in many respects.**

How can such channels be created? **By "institutionalizing" the American Black Press Establishment and making positively sure that they are intimately involved in any major activity taking place within the national black**

Where Is the Black Press?

community. My observations reveal that the black press seems to be scantily included or perhaps not included at all in such endeavors—until after the fact. In fact, it sometimes appears as though the white press has more of a profound presence on issues effecting blacks than do the black press. In other words, the points of origination of "significant black issues" seem to start with the white press, rather than the black press.

Launching of Pan Hel, Inc., must include the direct involvement of the black press, from the very beginning. Therefore, right here and now, I appeal to the black press/black media for their inclusion in launching the mass capital accumulation project discussed in Chapter Five. By black press/black media, I include newspapers, magazines, radio stations, TV stations, book publishers, journals, newsletters, tabloids, communications specialists via the computer or other electronic means, and other black run organizations that may publish information on blacks in general, such as think tanks, research centers, and etc.

For the same reasons that we cannot rely on white banks to fund black business development, we cannot rely on the white press/media to actively and accurately distribute vital wealth building information to the black community on a regular basis. This charge belongs to blacks. And, shaping up the black press for this and other challenges is every black man's and woman's responsibility. In effect, we must first support the black press; then tell them what we want. Thus, without question, among the first calls from the National Business League to launch Pan Hel, Inc. should be calls to the heads of national black press/media organizations.

On the other hand, why isn't the black press as aggressive and assertive as it once was? Sometimes I think the black press does not recognize its own potential for power. Does it not understand how much of a real influence it could have on the masses of blacks? **If anyone reads the black press, it is the black masses.**

So I have a few sole searching questions of my own. For example, several more church-affiliated black colleges will soon fail. All black colleges experience a dearth of respectable alumni support. This is information I did not first become acquainted with in the black press. It did not overlook the generosity of Oprah Winfrey, Bill Cosby, and Dr. Sam Nabrit to black colleges, **but it did not investigate the percent of the endowments that represents alumni giving. When will the black press take the lead in informing the American Black Community that our schools disproportionately subsist on charity from public and Establishment grants because of the lack of alumni support?**

The image of American Black Community is seriously pommelled in such instances. "As much as the least literate among us," the black press could chide, "the Talented Tenth cultivates the welfare mentality that some say characterizes American blacks." The black press was conspicuously silent when self-styled pipers fluted to America's unsophisticated, gullible and vulnerable poor. Did the press call for public debate on the Negro-colored-black issues, on the black/African-American issue or from the traditional and culturally conventional dress to the cultural intrusive garb of somewhat "alien" cultures.

Was its readership editorially advised to ask questions? In what direction and to what end would adopting such changes lead? What racial benefits would inure from indulging these changes? Will adopting them get the American Black Community attention or long-sought and much-needed recognition?

Where Is the Black Press?

Why such questions? Blacks have more compelling ends than new racial designations or jewel-adorned ears and noses. Our neighborhoods need to be restored to a level of civility where families stroll and lawn parties can be indulged without fear for our safety. Am I immune from danger as an **African-American** because I am attired in some strange tribal garb? Is that garb symbolic of an economic leap forward or academic enlightenment? Black press delinquency allowed Malcolm X to be upon us before 90 percent of us had ever heard of T. Thomas Fortune, **the godfather of black militancy.**

The sooner we shed the asinine notion that new names and fanciful slogans will propel the American Black Community economically forward and understand that only hard sacrifices such as liberal investments of our money and time in constructive and developmental projects will allow us to attain universal growth, the better. We will have come full circle back to the feet of Washington who taught those great lessons to transform his people from subserviency to competitiveness; lessons they could live by and on which they could build.

I was recently told that the noise from the echo chamber that sounds like repeats of new labels and fanciful slogans might include noise from the black press reaching for a pitch it has been told to reach, the equivalent of having been told to jump and its asking, "How high?" In short, the black press is in lock step on such matters. So the impending crisis in the black college community won't be reported first by the black media unless **aggressive journalism defines them**.

Thus, in view of the foregoing, our most crying need is a relevant black media, committed to self help and lifelong issues affecting us which have no predispositions of the invincibility and indispensability of high profiled blacks. A relevant black press would have warned the American Black Community in the 1960's that uttering "Black Power" and raising a clenched hand was more a cry for help than anything else; and that the Booker T. Washington principle of building a substantial business base and from there attempt to seize power is still valid.

As a child in New Orleans, I still remember legendary Rampart Street with several hundred businesses but just ten of them black-owned. The neighborhood grocery stores were owned by mostly whites of Italian and Greek decent, who had one thing in common with those of Asian extraction who are the owners of most of the neighborhood stores today: they spoke English with a thick accent.

Just as urgently needed is a black media that keep their black subscribers informed on vital matters: 35 million American blacks earn a total of about $360 billion yearly, a figure that will rise to about $400 billion in 1995. But, the total receipts for American's 450,000 black businesses in 1994 will be about $27 billion, roughly 7.5 percent of the American Black Community's gross annual income. That's the most profound statement of how well self-help **is not doing** in black America.

It's also the best argument in support of a national black monument in Washington, D.C. by and for blacks with black money only to put the all-important "we" in black strivings. Just as urgently needed too is a black media which prod blacks on the stage to be applauded—not entertained; to teach—not to recite grievances.

My fraternity sponsors a Future Leadership Club for 40-50 high school boys each year. They learn basic leadership by holding various offices in the club, whose motto is, "self-esteem, self-awareness and building responsibility."

Where Is the Black Press?

Baseball legend Henry Aaron shocked the boys one day when he told them sports superstars cannot be role models for them. They can be models, he told them, but models of success, like financial success.

Aaron told the boys, "Your role model should be someone you know better and see more often, like a neighbor, a relative, your teacher, doctor or class leader. When have you seen Deion Sanders last? When will you see me again? Are my values compatible with yours? But that does not mean you should not want to be successful like local Atlanta businessmen Gregory Baranco and Herman J. Russell, as well as superstars like Beau Jackson, David Justice, Charles Barkely, Randall Cunningham and others. Maybe when you get in college, they could become your role models, but not now. It's a big jump from here to where they are. Dream but know the difference between dreaming and fantasizing."

Aaron told me he had been saying that same thing to youth groups all over the country. I wonder why I hadn't read it in the black press or heard it on a broadcast or other medium. Aaron's comments constituted the greatest tribute to local heroes since Washington's epic observation that there is as much dignity in tilling the fields as in writing a poem.

Now is the time for the black press to again rise to its ultimate level of authenticity and influence. Perhaps all uniting behind the Pan Hel concept could be a progressive start.

Epilogue

The period 1787 to 1865 defines the white American population; the period 1865 to 1964 affirms that definition. Now, another definition must be found for the period since 1964 for white Americans. Let me explain a bit further.

The great human slaughters cited in some of the preceding chapters were intended to dispel the notion of hero(es) at every turn in history, not to lull blacks into believing the 25,000 blacks lynched since 1865 were comparatively insignificant. In fact, the millions of Chinese, Russians, Jews, etc., slaughtered are overshadowed by the cruelty, inhumanness, and heinousness that went into the lynchings to make them among the most viscous, savage, and barbarous acts of violence in civilized history. The blood of the millions cited stained the hands and the historical image of the few tyrants who ordered the massacres, not the people they ruled.

But no tyrannical ruler ordered approval of the U.S. Constitution which did not recognize blacks as citizens but that regarded blacks as property or economic goods; nor did a tyrant order the Civil Rights Act of 1875 to be de-

Epilogue

clared unconstitutional which cleared the way for Negroes to be denied the freedom, equal rights, benefits and protection guaranteed them as U.S. citizens under the amended U.S. Constitution. So the 25,000 black lynchings were not ordered by a tyrannical ruler, but were committed, sanctioned, encouraged and even applauded by the American white population whose hands and history are stained with the blood of those who were lynched, whose businesses were destroyed and who were otherwise maligned.

Since 1864, lynching of blacks have been taken over by a new entity—The Establishment—using a method that required no telltale bloodletting. The new method long ago received the endorsement and wholehearted approval of the white American population.

Unexpected bonuses have inured to The Establishment under the new method of lynching: The collective black income fell to just 60% of parity; the number of blacks turning to crime, ornamental gewgaws and superficial things increased; a sharp rise in the black male high school dropout rate occurred; a precipitous decline in the number of black males entering college (in a society that rewards education) came about; a high incidence of drug use among blacks evolved; an increase in the number of single parent homes developed; and, a Republican controlled 104th U.S. Congress came about which some blacks view as calamitous.

The above scenario is an encapsulated damage assessment of harm to black Americans resulting from "white crimes." It provides a background for sociologists, social scientists, historians and others to reformulate or reassess their perceptions of American history and maybe even rewrite it. It is a platform from which the National Black Bar Association could launch a program of legal aggressiveness that would convert its annual convention into an event of universal news worthiness; and it is the basis for insisting on, indeed, demanding participation of the majority business establishment and the American white community in the American Black Community's self help initiative.

This book contemplates a National Black Administrative Complex in Washington, D.C., paid for entirely by black contributions in which a great national black monument will be the centerpiece. The complex will include an Administrative building, an Enterprise building, The National Black Foundation Tower, a black population counter and a quality gauge that measures the present number of blacks with degrees versus the number with degrees one year earlier.

The goal for the foundation will be at least $40 billion in its first ten years of existence. The foundation money will be used to assure that no black student drops out of college because the tuition is not affordable; to establish and maintain remedial learning centers staffed by students receiving tuition assistance from the Foundation and qualified voters from the community; to support the College Bound Youth Program, etc. Management and supervision of such activities would perhaps become the responsibility of one of the conventional black organizations now in existence.

Also, the foundation's money will be used to construct a universal black cultural center sited near the geographic center of black America—wherever it may be—housing many museums and diverse pavilions operated by foreign governments, and where world class domestic and international competitive sports events will be held. It will house zoos and animal farms and be a place for the family and group vacations for recreation, relaxation and indulgence of their culture and heritage and meet the rest of the world.

Epilogue

The foundation will also be a source of a measure of funding for black oriented qualified organizations.

The goal of this initiative was circumscribed by rephrasing a passage from Washington's Atlanta speech: Over the first fifteen years to shave black responsibility to no more than 12% of the nation's crime and 10% of it's ignorance; to produce 12% to 15% or more of its economic, industrial and technological prosperity; own at least an equitable portion of its wealth and represent at least 15% of the nation's intelligence.

Those of the white population and The Establishment who answer our call to participate in this proactive and cooperative venture will be literally aiding black America in lifting a burden arbitrarily placed on it years ago. Those who ignore the call to participate will be figuratively pressing down the load to make it heavier and thus be deemed as favoring continuance of the lynching of black Americans.

But like the statement made by 50,000 Montgomery's Negroes in 1955 that resulted in the social restructuring of America, this book posits that a statement by America's 35 million blacks is an idea whose time has arrived, and that the power of black resolve to make that statement will in time reduce all roars of opposition to purrs of submission.

The ideal payoff for my efforts would be the leadership necessary to restore the National Business League to full-time operation to come from within the black business establishment to expedite this great black push in motion. This would effectively activate a series of messages of appreciation: from present day black America (35 million people) to Montgomery's 1955 black community of 50,000—to A. Phillip Randolf of 1941 and his poised-to-march 100,000 blacks—to Marcus Garvey and the 1920 black community of 12 million. That mutually-admiring quartet would in turn send a joint message to Booker T. Washington that would read: "Thanks for teaching us; your strategy works."

In conclusion, I have mostly described the rewards for being organized. Nothing precedes reactivating the National Business League and total organization of America's black businesses, in that order. Nor should the economic implication of these activities be taken lightly. The great **push** will **pull** many new businesses into existence, raise black employment levels and arouse in many latent black businesses talents and ambitions.

Black America has a deficit of over 50,000 lawyers, 60,000 medical doctors and dentists, 5 million degree bearing blacks and over 1 million businesses. A $40 billion plus foundation trove will afford some nice economic damage control. And power commensurate with our numbers will be much closer to reality.

—APPENDIX—

Booker T. Washington's Cotton States Exposition Speech, September 18, 1895

Mr. President and Gentlemen of the Board of Directors and citizens.

One-third of the population of the South is of the Negro race. No enterprise seeking the material, civil, or moral welfare of this nation can disregard this element of our population and reach highest success. I convey to you, Mr. President and directors, the sentiment of the masses of my race when I say that in no way have the value and manhood of the American Negro been more fittingly and generously recognized than by the managers of this magnificent exposition at every stage of its progress. It is a recognition that will do more to center the friendship of our two races than any occurrence since the dawn of our freedom.

Not only this, but the opportunity here afforded will awaken among us a new era of industrial progress. Ignorant and inexperienced, it is not strange that in the first years of our new life we began at the top instead of at the bottom; that a seat in Congress or the state legislature was more sought than real estate or industrial skill; that the political convention of stump speaking had more attraction than starting a dairy farm or truck garden.

A ship lost at sea for many day suddenly sighted a friendly vessel. From the mast of the unfortunate vessel was seen a signal, "Water, water, we die of thirst!" The answer from the friendly vessel at once came back, "Cast down your bucket where you are." A second time the signal, "Water, water; send us water!" ran up from the distressed vessel and was answered, "Cast down your bucket where you are." And a third and fourth signal for water was answered, "Cast down your bucket where you are." The captain of the distressed vessel, at last heeding the injunction, cast down his bucket, and it came up full of fresh, sparkling water from the mouth of the Amazon River.

To those of my race who depend on bettering their condition in a foreign land or who underestimate the importance of cultivating friendly relations with the southern white man, who is their next-door neighbor, I would say, "Cast down your bucket where you are." Cast it down in making friends in every manly way of the people of all races by whom we are surrounded.

Cast it down in agriculture, mechanics, in commerce, in domestic service, and in professions. And in this connection it is well to bear in mind that whatever sins the South may be called to bear, when it comes to business, pure and simple, it is in the South that the Negro is given a man's chance in the commercial world, and in nothing is this exposition more eloquent than in emphasizing this chance. Our greatest danger is that in the great leap from slavery to freedom we may overlook the fact that the masses of us are to live by the production of our hands, and fail to keep in mind that we shall prosper in proportion as we

Booker T. Washington's Cotton States Exposition Speech

learn to dignify and glorify common labor and put brains and skills into the common occupation of life; shall prosper in proportion as we learn to draw the line between the superficial and the substantial, the ornamental gewgaws of life and the useful. No race can prosper till it learns that there is as much dignity in tilling a field as in writing a poem. It is at the bottom of life we must begin, and not at the top. Nor should we let our grievances overshadow our opportunities.

To those of the white race who look to the incoming of those of foreign birth and strange tongue and habits for the prosperity of the South, were I permitted I would repeat what I say to my own race, "Cast down your bucket where you are." Cast it down among the eight millions of Negroes whose habits you know, whose fidelity and love you have tested in days when to have proved treacherous meant the ruin of your firesides. Cast down your bucket among these people who have, without strikes and labor wars, tilled your fields, cleared your forests, built your railroads and cities, and brought forth treasures from the bowels of the earth, and helped make possible this magnificent representation of the progress of the South.

Casting down your bucket among my people, helping and encouraging them as you are doing on these grounds, and to education of head, hand, and heart, you will find that they will buy your surplus land, make blossom the waste places of your fields, and run your factories. While doing this, you can be sure in the future, as in the past, that you and your families will be surrounded by the most patient, faithful, law-abiding and unrestful people that the world has seen. As we have proved our loyalty to you in the past, in nursing your children, watching by the sickbed of your mothers and fathers, and often following them with tear-dimmed eyes to their graves, so in the future, in our humble way, we shall stand by you with a devotion that no foreigner can approach, ready to lay down our lives, if need be, in defense of yours, interlacing our industrial, commercial, civil, and religious life with yours in ways that shall make the interests of both races one.

In all things that are purely social we can be as separate as the fingers, yet one as the hand in all things essential to mutual progress.

There is no defense or security for any of us except in the highest intelligence and development of all. If anywhere there are efforts tending to curtail the fullest growth of the Negro, let these efforts be turned into stimulating, encouraging, and making him into the most useful and intelligent citizen. Effort or means so invested will pay a thousand percent interest. These efforts will be twice blessed - "Blessing him that gives and him that takes."

There is no escape through law of man or God from the inevitable: -

The law of changeless justice bind
Oppressor with oppressed;
And close as sin and suffering joined,
We march to fate abreast.

Nearly sixteen millions of hands will aid you in pulling the load upward, or they will pull against you the load downward. We shall constitute one-third and more of the ignorance and crime of the South, or one-third to the business and industrial prosperity of the South, or we shall prove a veritable body of death, stagnating, depressing, retarding every effort to advance the body politic.

Gentlemen of the exposition, as we present to you our humble effort at an exhibition of our progress, you must not expect overmuch. Starting thirty years ago with own-

Booker T. Washington's Cotton States Exposition Speech

ership here and there in a few quilts and pumpkins and chickens (gathered from miscellaneous sources), remember the path that has led from these to the inventions and production of agricultural implements, buggies, steam-engines, newspapers, books, statuary, carvings, paintings, the management of drugstores and banks, has not been trodden without contact with thorns and thistles. While we take pride in what we exhibit as a result of our independent efforts, we do not for a moment forget that our part in this exhibition would fall far short of your expectations but for the constant help that has come to our educational life, not only from the Southern states, but especially from Northern philanthropists, who have made their gifts a constant stream of blessing and encouragement.

The wisest among my race understand that the agitation of questions of social equality is the extremist folly, and that progress in the enjoyment of all the privileges that will come to us must be the result of severe and constant struggle rather than of artificial forcing. No race that has anything to contribute to the markets of the world is long in any degree ostracized. It is important and right that all privileges of law be ours, but it is vastly more important that we be prepared for the exercise of these privileges. The opportunity to earn a dollar in a factory just now is worth infinitely more than to spend a dollar in an opera-house.

In conclusion, may I repeat that nothing in thirty years has given us more hope and encouragement, and drawn us so near to you of the white race, as this opportunity offered by the exposition; and here bending, as it were, over the alter that represents the results of the struggle of your race and mine, both starting practically empty-handed three decades ago, I pledge that in your effort to work out the great and intricate problem which God has laid at the doors of the South, you shall have at all times the patient, sympathetic help of my race; only let this be constantly in mind, that, while from representations in these buildings of the produce of field, of forest, of mine, of factory, letters, and art, much good will come, yet far above and beyond material benefits will be that higher good, that, let us pray God, will come, in a blotting out of sectional differences and racial animosities and suspicions, in a determination to administer absolute justice, in a willing obedience of all classes to the mandates of law. This, coupled with our material prosperity, will bring into our beloved South a new heaven and a new earth.

Select Bibliography

The following bibliography is not a complete listing of all of the material and sources I referenced in the development of this book. The books I have read span more than thirty years ever since I developed an interest in the history of blacks in America; the total list would include hundreds. Therefore, I am only providing a selected list that I am certain bear directly on the focus of this book.

Bennet, Lerone, Jr. *Black Power,* U.S.A. Chicago, 1967.
Bond, Horace Mann. *The Education of the Negro in the American Social Order.* New York, 1934.
Brawley, Benjamin G. *Negro Builders and Heroes.* Chapel Hill, 1937.
Brazeal, B.R. *The Brotherhood of Sleeping Car Porters.* New York, 1946.
Brisbane, Robert H. *The Black Vanguard.* Valley Forge, 1970.
Broderick, Francis L. *W.E. B. DuBois, Negro Leader in a Time of Crisis.* Stanford, 1959.
Carmichael, Stokley and Hamilton, Charles V. *Black Power.* New York, 1967.
--- . *The Politics of Liberation in America.* New York, 1963.
Cross, Theodore. *Black Capitalism.* New York, 1969.
--- . *The Black Power Imperative.* New York, 1987.
Douglass, Frederick, *The Life and Times of Frederick Douglass.* Hartford, 1881.
DuBois, W.E.B. *The Autobiography of W.E. B. DuBois.* Edited by Herbert Aptheker. New York, 1966.
Ellison, Ralph. *The Invisible Man.* New York, 1952.

Franklin, John Hope. *Reconstruction after the Civil War.* Chicago, 1962.

---. *From Slavery to Freedom.* New York, 1980.

Franklin, Robert M. *Liberating Visions.* Minneapolis, 1990.

Frazier, E. Franklin. *Black Bourgeoisie.* New York, 1962.

Gaston, A.G. *Green Power.* Birmingham, 1968.

Harlan, Louis R. *Booker T. Washington.* New York, 1972.

Higginbotham, Leon. *In the Matter of Color.* New York, 1978.

Hughes, Langston and Meltzer, Milton. *A Pictorial History of the Negro in America.* New York, 1968.

Lacy, Dan. *The White Use of Blacks in America.* New York, 1972.

Lewis, David Levering. *W.E.B. DuBois, Biography of a Race, 1869-1919.* New York, 1993.

Lincoln, C. Eric. *The Black Muslims in America.* Boston, 1961.

Literary Classics of the United States, Inc. *DuBois.* New York, 1986.

Logan, Rayford. *What the Negro Wants.* Chapel Hill, 1944.

Malcom X. *The Autobiography of Malcolm X.* Edited by Alex Haley. New York, 1964.

Marcus Garvey. *Marcus Garvey and Visions of Africa.* Edited by John Clark. New York, 1963.

Oak, Vishnu V. *The Negro's Adventures in General Business.* Yellow Springs, 1949.

Quarles, Benjamin. *The Negro in the Making of America.* New York, 1964.

Spencer, Samuel R. *Booker T. Washington and the Negro's Place in American Life.* Boston, 1965.

Thornbrough, Emma Lou. *T. Thomas fortune, Militant Journalist.* Chicago, 1972.

Washington, Booker T. *Up From Slavery.* New York, 1901.

---. *The Negro in Business.* Boston. 1907.

Select Bibliography

---. *The Story of the Negro*. London, 1909.
White, Walter. *A Rising Wind*. New York, 1945.
Woodson, Carter G. *The Mis-Education of the Negro*. Washington, 1933.
Young, Whitney M. Jr. *To Be Equal*. New York, 1964.

Index

A

Aaron, Henry 127
"African American," as race designation 25
African slave trade 30
"Afro-American," as racial designation 79
Afro-American League 38, 68
   as first civil rights organization 81
"Afro-Americans" as first suggested by T. Thomas Fortune 24
Aiken, South Carolina 43
   whites killing of blacks 37
Albert Wicker Junior High School 58
Alsop, Joseph 78
America Black Community strategy for advancement 69–70
   finding out its problems 24–25
   gross income 14
   potential for economic power 112
   state of black businesses 13
   steps toward organizing 31–32
American Black Community Foundation, 71
American black history
   as revisionist history 9–10
American white populace
   sanctioning of black lynchings 129
America's corporate empire builders 18
Andrew Johnson 35
Arab entrepreneurs 28
Arabs 17
Atlanta Constitution 38
"Atlanta Cotton States Exposition Speech of 1895" 9, 39, 61, 67, 133-138
Atlanta Life Insurance Company 110
Atlanta massacres 46
   and destruction of black businesses by whites 46

B

Baranco, Gregory 127
Barkely, Charles 127
Bennett, Lerone 23
Bethel Historical Society, 59
Bethune-Cookman College 15
"Black," as race designation 25
Black America
   annual income 115
   state of economic irrelevancy 116
Black America's deficits 132
Black Business Establishment 11
   as a foundation of power 109–121
   catalyst for change in the black community 17
   tripling in size 16
Black business organization
   as a lobby 109–110
Black businesses
   as targets of white rioters in Atlanta 110
   mass capital accumulation plan 112

Index

early growth of 46
first destruction of by whites 46
gross annual receipts 72
growth of attributable to Booker T. Washington 46
gross receipts vs. blacks' annual income 126
how many more are needed 16
number of in 1915 15
number of today 16
number people employed 115
organization of for advancement of blacks 11, 69–70
Black capitalism 114
and power 114–115
Black capitalism program
and blacks contempt of 91
Black Capitalism program, of the Nixon era 114
Black college graduates 14
Black cultural heritage
damage of 106
Black economic boom
golden age of 94
Black elected officials
as early republicans 40
"Black Enterprise 100" 96, 119
"Black Enterprise 1,000" 96, 119
"Black Enterprise 500" 119
Black Enterprise magazine 29, 105
Black entrepreneurs
during the early economic boom 96
Black history
as being faulty 27
as revisionist history 33–34, 53-55
maintaining image of 33
Black leaders/leadership

and self centeredness 85
challenge of with respect to NBL 48
challenge to determine what Washington's motives were 48
Black movie makers
issue concerning filming of black massacres 39
"Black Power" 90, 126
Black Press 122–127
importance of 122
need for in black economic movement 122
lack of aggressiveness 123–125
need to support 123
vs the white press 123
Black protest, in the South
as seen by Washington 60–61
Black slavery, end of 34
effects on Southern whites 34–35
Black Slavery War (or Civil War) 54 *See also* Issue of Black Slavery War
Black votes
importance of in 1800s 36
Black youth, problems of 96
Blanchet, Jr., Louis 59
Bonding symbol
for blacks 18
Booker T. Washington and the Negro's Place in American Life 69, 83
"Booker T. Washington: An Uncommon Perspective" 51
Boule Journal, 10, 51
Boyer, Joseph 59
Brown vs. Board of Education 103

Brunis, John Picket 59
Buddhism 86
Bunche, Ralph 112
Burns, Leonard 59
Burrell, Berkely G. 108
Business establishment, in relation to power 8
Business loans to blacks 114

C

Canoon, Mr. 20
Capital accumulation project for black businesses
  and the black press 123–124
  benefits of 112–113
Carpetbaggers 43
Carver, George Washington 13
Chamber of Commerce 72
Christianity 17, 86
  and black culture 102
Civil rights 91
Civil Rights Act of 1875 36, 46, 79, 128
  declared unconstitutional 37
  effects of overturning 37
  fate of under Hayes administration 37
Civil Rights Act of 1964 22, 90
  as King's triumph 28
Civil War (or Issue of Black Slavery War) 34 See also Black Slavery War
Clinton, President William requested to address NBL 47
Colfax black massacres 39
"Colored", as racial designation 78
Common goals for black America 105
Confederate states 36

Coushatta black massacres 39
Crisis magazine 81, 84, 89, 104
  as run by DuBois 84
Cross and Star of David 18
Culturally endemic bond 17
Cunningham, Randall 127

D

"Darkie" jokes 56–57
  as relates to Washington 59
"Darkie roles and mannerisms" show appearances of black entertainers 60
Delta Sigma Theta sorority movie made 116
Deusen, John C. Van 14
Dominant heroic trio 11
Douglass, Frederick 35, 39, 54, 92
  early motives for uplifting blacks 35
Dred Scott 22, 34, 38, 68, 79
DuBois, W.E.B. 9, 77, 79
  after break with Washington 99–100; after denouncing Washington 82; and the *Crisis* magazine 81; as a miscast 99; as founder of NAACP 10, 80; claim of changing attitude of the American Negro toward caste 87; denouncement of Washington's 1895 speech 86; disagreement with Washington 77; early view of Washington 81; legacy of 78; making the "color line" statement 79; opposition to the federal election bill 79;

Index

place in history 88; possibility of being used underhandedly by whites 100; reason brought to NAACP 101; Talented Tenth essay 14; view of Marcus Garvey project 104
Dunbar, Paul Lawrence 56
Dusk of Dawn 87

E

Ebony Businessmen's League 55–56
Ebony magazine's, 50 most influential blacks 34
Economic base, in relation to power 8
Ego builders 19
Ellison, Ralph 59
Empire builders 19
Establishment, The 115, 129
 practicing a new form of "lynching" 129
Establishment-Induced Black Oppression Centers (EIBOCs) 39, 86
Europeans 17
Executive Order 8802 68

F

Fortune, T. Thomas 9, 24, 61, 68, 74, 78, 89
 and Afro-American League 38
 and the black press 125
 conceives civil rights organization 91
 militant black activist and publisher 23
"Forty Acres and a Mule" 54
Fox, Redd 56

Franklin, John Hope 23
Frazier, E. Franklin 23
Free mulatto Negroes 80
Freedmen's Bureau 44
Full personhood 11, 97–99

G

Garvey, Marcus
 crusade, the 45
 fall of 94
 industrial program 92–93
 reason for success 93
Gaston, A.G. 119
Ghandi, Mahatma 89
Golden Calf 113, 115, 121
 concept of racial cooperation pertaining to black businesses 113–121
Grady, Henry 38, 78
Grant, Ulysses S. 36
Gray, Noel 59
Greenberg, Jack 103

H

Hamsburg, South Carolina 43
 whites killing of black militiamen 37
Harding, Joe 59
Harlan, Justice John M. 68
Hartford's black business activists, 45
Hartford's Black Business Establishment slogan 73
Hayes, Rutherford, B. 36
Henderson, Alexa 110
Herndon, Alonzo F. 119
Historical worthiness of leaders defined 54–55
History-worthiness, measuring of

10
Houston, Charles 103
Hughes, Langston 24

I

"I Have A Dream" speech 28
Indispensable Dominant Heroic Trios 35, 39
   third trio 46
Industrial education for blacks as taught in earlier years 57–58
Islam 17
Issue of Black Slavery War (or the Civil War) 34, 35, 44, 45

J

Jackson, Andrew 22
Jackson, Beau 127
Japan
   indebtedness to black community 29
Jefferson, Thomas 35
Jews
   endemic bond 17
   friction between blacks 102
Johnson, Andrew President
   granting of amnesty to white males who killed blacks 36
   opposed compensation or concessions of any kind to blacks 35
Johnson, James Weldon 92
Johnson, Rosemond 92
Joseph A. Craig Elementary School 58
Judaism, 17
Justice, David 127

K

Kajiyama, Serioku
   comments regarding black Americans 29
King, Martin Luther Jr. 9, 19, 22, 25, 35,
   trouble finding funds for movement 71, 94
   demonstrating for rights, not power 90–91
   legacy 27
   monetary impact of his contributions 72

L

Leaders, responsibility of 24
Leadership crisis 104–105
Leland College 15
"Lessons learned" conferences
   need for 104
Lewis, David Levering 80, 83, 101
"Lift Every Voice and Sing" 10, 92
Lincoln, Abraham 35
Lipmann, Walter 78
Louisiana Constitution Convention 61
Louisiana
   the black killing grounds 39

M

Madison, James 35
Malcolm X 9, 89
   and the black press 125
Markham, Pigmeat 56
Marshall, Thurgood 35, 103, 112,

Index

Martial law
  reasons for 43–44
Martin Luther King crusade, the reason for success 93
Martinez, Mr. 58
Masons, Detroit
  TV station created 116
Mass dynamics 70
Mays, Benjamin 88
McDonogh No. 35 Senior High School 58
McKay, Claude 59
Memphis 43
Minorities 21
Montgomery Movement 9
Morehouse Medical School 15
Morgan, J.P. 38
Moton, Robert R. 88
Muhammadanism 86
Mulzac, Hugh 24

N

NAACP 75, 81, 91, 100
  creation of 80–81
  organization of 38
  perception of by Washington and others 92
NAACP historians
  questions to ask themselves 75
Nakasone, Yasushiro, 29
National Afro-American Council 74, 75, 81, 81–82
National Afro-American Council Convention 81
National Afro-American League 61, 74
National Black Administrative Complex 71, 130
National Black Enterprise building 130
National Black Foundation 130
  purposes and benefits of funding levels 130
National Black Monument 11, 17, 71, 126, 130
National Business League 11, 15, 47, 72, 91, 108, 123
  as facilitator for black reparations 31
  as part of black heritage 109
  as the administrative center of the black business 70–71
  catalyst for black economic movement 17
  creation of 82
  early objectives 18
  friction between DuBois and Washington about league 82–83
  goals of, with the black community 15
  organization of 15, 46
  purpose and formation on in early days 15
  purposes of—ala Washington 55
  Restoration and Preservation Committee 11
  restoration to full capacity 131
  support of 95
  use of to leverage black community 48
National Colored League 74
National Education Association Convention 69
National Negro Conference 80
National Urban League 75
  Washington's view of 75
"Negro," as racial designation 78

Negro New Englanders 74
Negro Reconstruction 78
New England Society of New
 York 38
New Orleans 43
New Orleans black massacres 39
New York Times 38
Newman, Joe 59
Niagara Conference 83
Niagara Movement 81
Non-blacks with businesses in
 black communities 47–48
Norris, Mr. 58

O

Oak, Vishnu 94
"On Mr. Booker T. Washington
 And Others," 77, 82
 criticism of Washington 101
Opelousas black massacres 36,
 39, 43
Organization, of the black
 community
 pre-condition for 17
Ovington 80
Ovington, Mary 101

P

Paine, Thomas 35
Pajeaud, Noella 59
Pan-Hel, Inc. 117–118, 123
Pan-Hellenic (Greek-letter)
 societies 116–117
 as part of capital accumulation
  project 116
Parks, Rosa 54
Pegler, Westbrook 78, 90
Pichon, Jr, Walter 59
Plessy vs Ferguson 46, 53, 68, 79
Potok, Chaim *8*
Power, fighting for
 versus fighting for rights 90
Power, in relation to economic
 base 8
Power seekers 92
Pryor, T.M
 issues as leader and spokesman
  for Ebony Business League
  64

Q

Quarles, Benjamin 23

R

Race issue in the South 38
Racial goals for blacks 96–98
Rampart Street 126
Randolph, A. Philip *9*, 35, 39,
 57, 92
 and Executive Order 8802 23
Reconstituting the Union 42
Reparations for blacks 30 *See
 also* National Business
 League
Republican Party 43
 deception of blacks 44–45
Republican political strategy
 meetings 43
Revisionist black history 87
"Revolution" 87
Revolutionary War
 blacks fighting in 22
 blacks presence in 42
Rights agitators 92
Rights, fighting for
 versus fighting for power 90
Ritchie, Lionel 59

Index

Robeson, Paul 56
Roosevelt, Teddy 22
Roosevelt, President Franklin D. 23, 96
Root, Elihu 38
Root waterers, 26
Rowan, Carl 78
Rufus Saxon 35
Russell, Charles Edward
   and the NAACP 80
Russell, Herman J. 59, 127
Rutherford B. Hayes
   ending of guards at voting polls in the South 36
   ending of martial law in the South 36

S

S. S. Booker T. Washington 24
Sanders, Deion 127
Scalawags 43
Schuyler, George S. 78
Scottsboro Boys 103
Second Confiscation Act of 1862 35
Sigma Pi Phi Fraternity *10*
Southern white males 35
   genocidal campaign against blacks 36
Spencer, Jr. Samuel R. 69, 83
Spingarn, Joel 102
St. Cyr, Edna 58
Stevens, Thaddeus 54
Stowe, Harriet Beecher 35, 39

T

T. Thomas Fortune, Militant Journalist 24, 78
Talented Tenth 100

as advanced by DuBois 84
   worthiness of 85–86
Taney, Roger B. 79
The Age newspaper, 68, 89
The Black Man in White America 14
The Chosen, 8
The Guardian 101
"The Latest Color Line," 79
The Mis-Education of the Negro 57, 84, 89
The Negro's Adventure in General Businesses, 94
The Souls of Black Folk 77, 78, 81
Thornbrough, Emma Lou 24, 78, 89, 91
Tiffany, Charles 38
Trotter, William Monroe 101
   Washington's severest critic 74
Truman, President Harry 68
Tulsa Oklahoma massacres
   and destruction of black businesses 46
Tuskegee University 15
   impact on education of blacks 13
   impact on the South's farm economy 13
Tuskegee Normal School 20

U

U.S. Constitution
   and blacks 128
Universal Negro Improvement Association 93
Up From Slavery 8, 67, 73, 84

V

Vann, Robert 88
Villard, Oswald Garrison 38
   and the NAACP 80
Voting Rights Act 22

W

W.E.B. DuBois: Biography of a Race, 1868-1919: 80
Walker, Madame C. J. 119
Walling 80
Warren Court 22
Washington, Booker T. 8, 9, 90, 100
   abdication of the South 61; advocating industrial education 57; and the NAACP 73; as creator of national black business leadership 49; as a teacher of his race 61; as father of black self help 47–48; as founder of Tuskegee Univ. 13; as national black leader 49; as president of NBL 63; as principal of Tuskegee 20; as undersubscribed 10; counsels blacks to building businesses 75; damage of reputation by black intellectuals 54; dealing with hostile whites 45; denigration of 9; desire to have a black power business base 114; elected president of NBL 82; emphasis on building black businesses 44-45, 126; emphasizing first an economic foundation for blacks 67; his teachings 51–52; historical image 33; impact of 1895 Cotton States speech 48–50; impact on Garvey 96; industrial trades at Tuskegee 59; jealousy and rage of by others 53; knowledge of Republican Party 44; legacy of 77; lesson from autobiography *Up From Slavery* 8; most misunderstood statement 42; national reputation and his handling of 63; position of leadership 65; prescription for blacks 44; purposes of NBL 55; receiving Harvard's honorary Master of Arts degree 76; respect for individual choice 59; speech at NEA 37; suggestion of reason to be organized 112–113; testing of the Louisiana suffrage laws 62; view of National Urban League 75; warning to whites not to deny blacks certain right 45–46; worst fear 103
Washington-bashing 33
Washington's Cotton States Exposition Speech lessons of 39–50
Watanabe, Michiko comments regarding black Americans 29
Weaver, Robert C. 112
White American populace how defined in relation to blacks 128

White press
  reliance on 123
Will, George 78
William T. Sherman 35
Willis, Alonzo 93
Wilmington, N.C. black massa-
  cres 46
Wilson, Flip 56
Woodson, Carter G. 57, 84, 89

Y

Yasushiro Nakasone
  comment regarding black
    Americans 29
Young, Andrew 27
Young, P.B. 88